CHROMOSOMES in HUMAN CANCER

CHROMOSOMES in HUMAN CANCER

By

JAROSLAV CERVENKA, M.D., C.Sc.
*Associate Professor, Division of Oral Pathology
and Division of Human and Oral Genetics
School of Dentistry
Associate Professor, Graduate School
Department of Genetics, College of Biological Sciences
University of Minnesota, Minneapolis, Minnesota*

and

LUCIEN KOULISCHER, M.D., M.A. Agrégé de l'Enseignement Supérieur
*Chief, Department of Cytogenetics
Institut de Morphologie Pathologique
Loverval, Belgium*

Edited by

ROBERT J. GORLIN, D.D.S., M.S.
*Professor and Chairman, Division of Oral Pathology
School of Dentistry
Chief, Human Genetic Clinic, Health Science Center
University of Minnesota, Minneapolis, Minnesota*

RC262
C47
1973

CHARLES C THOMAS · PUBLISHER
Springfield · Illinois · U.S.A.

290321

Published and Distributed Throughout the World by
CHARLES C THOMAS · PUBLISHER
BANNERSTONE HOUSE
301-327 East Lawrence Avenue, Springfield, Illinois, U.S.A.

This book is protected by copyright. No part of it may be reproduced in any manner without written permission from the publisher.

© *1973, by* CHARLES C THOMAS · PUBLISHER
ISBN 0-398-02629-7
Library of Congress Catalog Card Number: 72-84140

With THOMAS BOOKS *careful attention is given to all details of manufacturing and design. It is the Publisher's desire to present books that are satisfactory as to their physical qualities and artistic possibilities and appropriate for their particular use.* THOMAS BOOKS *will be true to those laws of quality that assure a good name and good will.*

Printed in the United States of America
H-2

PREFACE

Among the manifold challenges of medical science, the etiology of cancer remains the most intriguing for many reasons. Our understanding of the problem has advanced *pari passu* with the many advances in virology, immunology, cell surface research, and an inordinate number of other fields of biomedical research. Cytogenetics in our opinion occupies an important place in this extensive list.

It has been realized that chromosomes are not only involved in transmission of genetic information from generation to generation and from a cell to its daughter cells, but they are also responsible for the ordered release of this information for control of cellular development and function. Data have accumulated concerning systems of gene control and chromosomal activation or inactivation in complex organisms. Quite recently there has been a flurry of investigation of heterochromatins and satellite DNA, not only on the molecular level, but on the level of chromosomal morphology and function.

It seems justified that this line of research has induced considerable enthusiasm among the cytogenetic community and has stimulated Dr. T. C. Hsu to introduce the 9th Annual Somatic Cell Genetics Conference in Galveston (1971) by reminiscing that, "After the great excitement of the late 1950's and early 1960's the field of human cytogenetics has become somewhat dull . . ." and expressing hope that the new staining techniques ". . . should stimulate some interest for at least a few more years."

At this point in time we have decided to gather all available information on chromosomes in human tumors with the fervent hope that they might be used as a source of understanding of the seemingly disordered chromosomal constitution in human cancer cells and of the few rules which may be derived from these data.

We are aware that our attempts come at a time which we consider to be the threshold of a new era of qualitative cytogenetics in man. We hope that our efforts may be exploited by workers involved in clinical research as well as in research of the cancer cell *in vitro*, despite the fact that we have omitted the subject of chromosomes in established cell lines and in cancers of experimental animals.

It is our sincere desire that this book will contribute to understanding the manifold problems involved in human cancer etiology being a comprehensive and comprehensible overview on the form of genetic material so commonly altered during malignant transformation. Furthermore, it is our desire that this survey may serve a heuristic role in stimulating others to examine these problems.

There is an ancient apothegm that a tome has its incipience in enthusiasm, is continued in anxiety, and reaches fruition in exhaustion. This effort has tested that rule and has found it to be true. After surveying the plethora of papers on this subject, we are effete.

Aiding us in our efforts and to whom we express profound gratitude are Dr. Claude Fievez, Hattie L. Thorn, Marie Anne Cosme, Roberta J. Lensink, Eliane and Christiane Canon, Suzan M. Schwarze, and Bridget A. Stellmacher.

We wish to express our special appreciation to Saša and Paulette.

JAROSLAV CERVENKA
LUCIEN KOULISCHER

CONTENTS

	Page
Preface	v

PART I
General Considerations

Chapter
- I. MECHANISM OF CHROMOSOMAL REARRANGEMENTS 5
- II. COMMON CHROMOSOMAL PATTERNS IN CANCER 14
- III. CONGENITAL CHROMOSOMAL ABNORMALITIES AND CANCER 19
- IV. VIRUSES AND CHEMICAL CLASTOGENS 32
- V. ANALOGY OF KARYOTYPE EVOLUTION IN CANCER AND IN MAMMALIAN SPECIATION 37
- VI. THE USE OF TECHNIQUES REVEALING CONSTITUTIVE HETEROCHROMATIN 46

PART II
Hematopoietic and Lymphatic Systems

- VII. CHRONIC MYELOID (GRANULOCYTIC) LEUKEMIA 55
- VIII. ACUTE LEUKEMIAS 78
- IX. MYELOPROLIFERATIVE DISORDERS 92
- X. NON-LEUKEMIC HEMATOLOGICAL DISORDERS: PRELEUKEMIAS, ANEMIAS, DYSPROTEINEMIAS 106
- XI. MALIGNANT DISEASES OF THE LYMPHORETICULAR SYSTEM 119

PART III
Solid Tumors

- XII. FEMALE REPRODUCTIVE SYSTEM 131
- XIII. MALE REPRODUCTIVE SYSTEM 149

XIV.	Alimentary Tract	154
XV.	Respiratory Tract	170
XVI.	Nervous System	176
XVII.	Urinary System	186
XVIII.	Miscellaneous Tumors	192

Index 199

CHROMOSOMES in HUMAN CANCER

PART I

GENERAL CONSIDERATIONS

Chapter I

MECHANISM OF CHROMOSOMAL REARRANGEMENTS

CHROMOSOME STUDIES in malignancy are, in many respects, quite different from cytogenetic analyses performed in individuals with congenital malformations. In the latter the karyotype remains stable from conception (or from early stages of development) throughout the life of the individual, while in malignancy, important karyotypic changes may occur during a relatively short lapse of time. The approach to chromosome studies in cancer is dynamic, much as the development of the malignant tumor itself with its onset, growth, and metastasis.

It is useful to keep in mind that any chromosome analysis performed on a tumor tissue reflects changes which are present only at a particular time and site, being but one increment in the progression of the malignant process. This is the reason why the cytogenetics of neoplasia may appear at first glance to be confusing. By analogy, it represents but a single photographic enlargement printed from a long cinematographic film. One of the tasks of the cytogeneticist is to interpret the observations in context and to understand the mechanism of transformation of the original karyotype. This chapter deals with some basic concepts of chromosome change in malignancy and with possible ways which may eventuate in an abnormal karyotype in a given tumor.

In 1931, Levine wrote the following:

> Comparatively no phase of the cancer problem has received less attention in the last decade than the cytological phenomena of cancer cells in division. Obviously, this deserves more attention. According to our present knowledge, cancer is directly bound up with the process of nuclear and cell division. Without cell division, there is no cancer.

Today, the reverse situation is actually true: during the last decade, a spectacular amount of work has been devoted to cell division, and especially to chromosome studies of cancer cells. This has been possible since new cytogenetic techniques have been introduced. These techniques have been used with success both in the study of congenital and malignant chromosome aberrations.

BASIC CONCEPTS

The significance of chromosome abnormalities in malignancy is classically attributed to Boveri (1914). Impressed by the work of von Hansen concerning multipolar mitoses in cancerous tissues (see Turpin and Lejeune, 1965), Boveri suggested that chromosomal change is the primary trigger mechanism in cancerogenesis. Winge (1930) considered the role of chromosome changes in cell selection within tumor tissue. As noted earlier, review of the subject "chromosomes and malignancy," was carried out as early as 1931 by Levine. Since then, numerous works have been devoted to chromosome alterations of spontaneous and induced tumors in animals.

These investigations furnished the basis for the important concept of the *stemline* (Levan and Hauschka, 1953; Makino, 1957; Hsu, 1961): (a) the stemline is mainly responsible for the growth of the tumor, (b) in some neoplasms, one or more stemlines can exist and develop, (c) the cells of a stemline are characterized by a common chromosome set, and (d) cells with other karyotypes represent "sidelines" having small effect on the progression of the tumor. However, depending upon environmental circumstances, a sideline may become one or more stemlines.

In some cases, a *marker chromosome* can be observed. A *marker chromosome* in a tumor is one with a peculiar morphology, easily identified and, as a rule, not seen in the normal karyotype. The presence of a marker in all cells of a malignant neoplasm gives support to the hypothesis of the clonal origin of the tumor: the marker has been transmitted to all tumor cells by a single mutant cell from which they derive (Berger, 1968). Indeed, cytogenetics could be used as well as the study of glucose-6-phosphate dehydrogenase (Linder and Gartler, 1965; Gartler *et al.*, 1966; Beutler *et al.*, 1967; Fialkow *et al.*, 1967) to demonstrate either the single cell or multiple cell origin of a tumor.

In man, reliable cytogenetic studies in malignancy have been possible only since the discovery of the correct number of human chromosomes by Tjio and Levan (1956). Progress in this field has been spectacular since the observation of a specific marker in chronic myeloid leukemia by Nowell and Hungerford in 1960.

POSSIBLE MECHANISMS OF CHROMOSOME CHANGES IN MALIGNANCY

Malignant stemlines are often characterized by abnormal chromosome numbers and/or the presence of markers. Some possible mechanisms leading to these karyotype changes are herein described. For a better understanding of the formation of nonspecific or unstable abnormalities such as breaks, gaps, sporadic rings, or isochromosomes, the reader is referred to Rieger et al. (1968).

Polyploidy

Polyploidy is defined as the state in which a cell shows three, four, five, or more complete chromosome sets instead of the normal two present in diploid somatic cells (Rieger et al., 1968). A polyploid cell arises from a normal diploid cell when DNA synthesis occurs before division and for reasons unknown, there is failure of the cell to divide. Polyploidy is very often observed in malignant tissues; however, it is also observed in normal tissues such as liver, endometrium, urinary bladder epithelium, megakaryocytes, and in tissue cultures of normal diploid somatic cells. In cancer, polyploid cells may sometimes compose the stemline.

Endoreduplication

This is a special form of polyploidization, taking place within an intact nuclear envelope (Levan and Hauschka, 1953). An endoreduplication is established with certainty only if it is followed by normal mitosis. The chromosomes are seen lying topographically adjacent to one another, two-by-two (or four-by-four). Endoreduplication can lead to high levels of polyploidization and may be responsible for those tumor cells containing several hundreds of chromosomes.

Hyperploidy

This form of aneuploidy is characterized by the presence of one or more supernumerary chromosomes or chromosome segments. Thus a typical hyperdiploid cell exhibits 47 or more chromosomes, a hypertriploid mitosis 70 chromosomes or more, etc. Hyperploidy may be arrived at through different mechanisms, most frequently being attributed to nondisjunction. At anaphase, the chromosome pair fails to separate and migrates towards the same pole. One daughter cell will thus have a supernumerary chromosome, while the other will lack a chromosome. The cause of nondisjunction is uncertain and has been attributed to aging, viruses, chemicals, or, for example, in maize, to the presence of a mutant gene called "sticky." Indeed, perfect spindle function is also important to avoid nondisjunction. *Selective endoreduplication* of only one or a few chromosomes (Lejeune *et al.*, 1966; de Grouchy *et al.*, 1967) is another possible avenue leading to hyperdiploidy. The nucleolus may be responsible for nondisjunction also, especially in chromosomes carrying the nucleolus organizers (Ohno *et al.*, 1961) since close association during interphase could favor nondisjunction during mitosis.

Hypoploidy

In this condition one or more chromosomes of the diploid set are missing. There are essentially two mechanisms involved. One is *nondisjunction* and the other is *chromosome lagging*. The latter represents loss of one (or more) chromosome during mitosis due to noninclusion in either daughter cell at telophase.

Pseudodiploidy and Marker Chromosomes

In some metaphase spreads, the expected diploid number of chromosomes can be observed. However, the karyotypes may show "odd" or marker chromosomes. Deletions, inversions, translocations, segmental duplications, insertions, formations of ring or isochromosomes can account for the presence of the markers. (A definition of these terms may be found in Rieger *et al.*, 1968.) Such a cell containing the diploid number 46, one or more chromosomes being abnormal, is called *pseudodiploid*. When a structural rearrangement is found in the majority of mitoses of a tumor, we assume that this characteristic stemline is responsible for the tumor propagation. An example of a specific marker in a tumor is the Philadelphia chromosome (a deleted G group chromosome) found in human chronic granulocytic (myeloid) leukemia. The discovery of a marker chromosome in a malignant tissue is very important since cells with different karyotypes but containing the same marker can be related to a common ancestral cell.

Combined Mechanisms of Aneuploidization

More than one mechanism can be involved in malignant karyotype changes, especially if only one observation has been made. For example, nondisjunction may result in a hyperdiploid cell; translocations may occur between several chromosomes resulting in formation of markers; if chromosome losses then occur (including some markers), the pathways leading to the stemline karyotype actually observed can be completely masked. As said earlier, any one particular observation corresponds to just one single frame of a long movie. Nevertheless, information obtained in such a manner may be useful and worthy of interpretation.

CHROMOSOME EVOLUTION

Serial observations of the same tumor in short time sequences may exhibit important variations of karyotype, thus demonstrating a high degree of chromosome instability of the malignant clone. In some cases these variations are markedly progressive and demonstrate a real "evolution" that has been compared to the evolution of species (Lejeune, 1965 and Chapter V) and described by the term *clonal evolution*. Eventually, the presence of a marker chromosome suggests the common origin of the karyotypes observed. Many examples of clonal evolution are now known (see Table I). One of the principal tasks of the cytogeneticist studying a single tumor is to reconstruct the development of its karyotype from the normal karyotype to obtain insight into the progress of malignant transformation.

Theoretically, any chromosomal change in the cell results in severe genetic imbalance. It is easy to understand that karyotype transformations cannot occur randomly—some genetic disturbances may prove lethal to the cell. Therefore, Lejeune (1965) proposed three "laws" of clonal evolution:

TABLE I
CLONAL EVOLUTION IN A CASE OF ACUTE LEUKEMIA

Chromosome Number	Karyotype
47	tris 21
48	tris 21 + 1G
49	tris 21 + 2G
50	tris 21 + 2G + 1D
51	tris 21 + 2G + 2D
52	tris 21 + 2G + 2D + 1C
53	tris 21 + 2G + 2D + 2C
54	tris 21 + 2G + 2D + 2C + 1F
55	tris 21 + 2G + 2D + 2C + 2F

Note: From Lejeune et al., 1963.

1. Clonal evolution is progressive: successive abnormalities finally lead to the observed karyotype of the tumor tissue studied.

2. There is a tendency to duplication of supernumerary chromosomes (Table I).

3. So-called *forbidden combinations* exist. A new karyotype implies a new genetic pattern for the cell. The action of genes carried by "new" chromosomes will most likely alter the cell metabolism. If new requirements cannot be satisfied, the cell will die: these are the forbidden combinations inconsistent with the development of the tumor. Another example of a forbidden combination is the one allowing the host to immunologically reject the tumor clone.

CHROMOSOMES AND PREMALIGNANCY

The cytogenetics of *premalignancy* is indeed interesting and offers an opportunity to better understand the relationship between chromosome aberrations and the origin of cancer. Are chromosomes normal in premalignant lesions? Do some peculiar changes signify transformation from benign to malignant cells and is cytogenetics a useful tool for early detection of cancer? The answers to these questions depend upon the definition of premalignancy.

The concept of premalignancy has stimulated many discussions. For Willis (1967), some tumors are to be considered premalignant since they often transform to malignant types. This is true for localized tumors of the bladder, Bowen's disease of the skin, and familial polyposis of the colon or rectum. Over 10 percent of patients with testicular tumors show ectopy of the testicle; the relationship between Paget's disease of the nipple and cancer is open to discussion, etc. (For detailed analysis see Lynch, 1967). Stewart's (1950) statement: "The female breast is a precancerous organ" is well-known. In the case of leukemia, many diseases have been called "preleukemic": aplastic anemia, sideroachrestic anemia and other anemias, and thrombocythemia. All myeloproliferative disorders that may transform into chronic myeloid leukemia (Dameshek and Gunz, 1964) are also "preleukemic," as is trisomy 21 syndrome with its high proclivity to develop acute leukemia (Krivit and Good, 1956, 1957; see also Chap. III).

The best cytogenetic approach to this problem seems to be serial chromosomal analysis of nonmalignant tissue, followed by subsequent observation of *the same tissue* when the disease has undergone malignant transformation. Little data based upon these criteria have been published. The most valuable are those concerning CML (chronic myeloid leukemia) because of the presence of a specific marker: the Ph^1 chromosome. In some cases (Kemp et al., 1961, 1964), the Ph^1 chromosome has been observed prior to the appearance of any actual signs of CML. In at least one case, polycythemia vera with Ph^1 chromosome exhibited transformation into CML, thus suggesting that chromosomal aberration occurred *before* clinical onset of the disease. Other cases with a Ph^1 chromosome, but without CML, have been reported in myeloproliferative diseases suggesting future onset of CML.

Less instructive are the data obtained in neoplastic disorders of the cervix uteri (see Chap. XII). Cytogenetic investigations have been made in cases of severe dyplasia of the cervix, carcinoma *in situ*, and invasive carcinoma (Wakonig-Vaartaja and Kirkland, 1965; Wakonig-Vaartaja and Hughes, 1967). The karyotype is, as expected, more frequently normal in dysplasia, and aneuploid (within a wide range) in malignant lesions of the cervix. However, aneuploidy is frequently found in dysplasia. When comparison is made with CML, it may be noted that even a relatively unimportant chromosome change (deletion of the long arms of one of the smallest chromosomes of the set) is sufficient for development of malignancy. In the case of cancer of the cervix, only a statistical analysis is meaningful and, in a single case, even considerable karyotype rearrangements cannot be interpreted as proof of malignancy.

Other examples from each system will be presented later. At this point, however, it must be stressed that severe aneuploidy with marker chromosomes can be observed in nonmalignant tumors and normal karyotypes can be observed in some cancers. These remarks do not tend to deny the importance of cytogenetics in malignancy, but suggest that interpretation of data must be made with considerable caution.

REFERENCES

Berger, R.: Sur la méthodologie de l'analyse des chromosomes des tumeurs. Thèse, Faculté des Sciences de Paris, 1968.

Beutler, E.; Collins, Z.; Irwin, L. E.: Value of genetic variants of glucose-6-phosphate-dehydrogenase in tracing the origin of malignant tumors. N Engl J Med, 276:389, 1967.

Dameshek, W., and Gunz, F.: Leukemia, 2nd ed. Grune and Stratton, New York and London, 1964.

Fialkow, P. J.; Gartler, S. M.; Yoshida, A.: Clonal origin of chronic myelocytic leukemia in man. Proc Nat Acad Sci USA, 58:1468, 1967.

Gartler, S. M.; Ziprkowski, L.; Krakowski, A.; Ezra, R.; Szeinberg, A.; Adam, A.: Glucose-6 phosphate-dehydrogenase mosaicism as a tracer in the study of hereditary multiple trichoepithelioma. Am J Hum Genet, 18:282, 1966.

Grouchy de, J.; De Nava, C.; Bilski-Pasquier, G.; Zittoun, R.; Bernadou, A.: Endoréduplication sélective d'un chromosome surnuméraire dans un cas de myélome multiple (maladie de Kahler). Ann Genet, 10:43, 1967.

Hsu, T. C.: Chromosomal evolution in cell populations. Int Rev Cytol, 12:69, 1961.

Kemp, N. H.; Stafford, J. L.; Tanner, R. K.: Chromosome studies during early and terminal chronic myeloid leukemia. Br J Med, 1:1010, 1964.

Krivit, W.; and Good, R. A.: The simultaneous occurrence of leukemia and mongolism. Am J Dis Child, 91:218, 1956.

Krivit, W., and Good, R. A.: Simultaneous occurrence of mongolism and leukemia. Am J Dis Child, 94:289, 1957.

Lejeune, J.: Leucémies et cancers. In Turpin, R., and Lejeune, J.: Les chromosome humains. Paris, Gauthiers-Villard, 1965.

Lejeune, J.; Berger, R.; Rethore, M. O.: Sur l'endoréduplication sélective de certains segments du génome. C R Acad Sci (Paris), 263:1880, 1966.

Levan, A., and Hauschka, T. S.: Endomitotic reduplication mechanisms in ascites tumors of the mouse. J Nat Cancer Inst, 14:1, 1953

Levine, M.: Studies in the cytology of cancer. Am J Cancer, 15:144, 1931.

Linder, D., and Gartler, S. M.: Glucose-6-phosphate-dehydrogenase mosaicism: utilization as a cell marker in the study of leiomyomas. Science, 150:67, 1965.

Lynch, H. T.: Hereditary factors in carcinoma. In Recent Results in Cancer Research. Heidelberg-New York, Springer-Verlag, 1967, vol. 12.

Makino, S.: The chromosome cytology of the ascites tumors of rats, with special reference to the concept of the stemline cell. Int Rev Cytol, 6:26, 1957.

Nowell, P. C., and Hungerford, D. A.: A minute chromosome in human granulocytic leukemia. *Science, 132*:1497, 1960.

Ohno, J.; Trujillo, J. M.; Kaplan, W. D.; Kinosita, R.: Nucleolus organisers in the causation of chromosomal anomalies in man. *Lancet, 2*:123, 1961.

Rieger, R.; Michaelis, A.; Green, M. M.: *A Glossary of Genetics and Cytogenetics*. Berlin-Heidelberg-New York, Springer-Verlag, 1968.

Stewart, F. W.: *Tumors of the Breast*. Armed Forces Institute of Pathology (Publishers), Washington, D. C., 1950.

Tjio, J. H., and Levan, A.: The chromosome number of man. *Hereditas, 42*:1, 1956.

Turpin, R., and Lejeune, J.: *Les chromosomes humains*. Gauthiers-Villars, Paris, 1965.

Wakonig-Vaartaja, R., and Kirkland, J. A.: A correlated chromosomal and histopathologic study of preinvasive lesions of the cervix. *Cancer, 18*:1101, 1965.

Wakonig-Vaartaja, R., and Hughes, D. T.: Chromosome studies in 36 gynaecological tumours of the cervix, corpus uteri, ovary, vagina and vulva. *Europ J Cancer, 3*:263, 1967.

Willis, R. A.: *Pathology of Tumors*, 4th ed. Butterworth, London, 1967.

Winge, O.: Zytologische Untersuchungen über die Natur maligner Tumoren. II. Teerkarzinome bei Mäussen. *Z Zellforsch, 10*:683, 1930.

Chapter II

COMMON CHROMOSOMAL PATTERNS IN CANCER

IT HAS BEEN OBSERVED that a markedly high frequency of chromosome aberrations occurs in human cancer. Aneuploidy is certainly one characteristic of malignancy and cytogenetics provides a suitable approach to better understanding of the malignant process at the cellular level. The purpose of the present chapter is to discuss the possible significance of the presence or absence of karyotype abnormalities in cancer cells. Discussion is complicated for at least two reasons: (a) we do not know what causes transformation of a normal cell into a malignant cell, and (b) tumors histologically similar may exhibit different chromosome abnormalities.

In general, chromosome changes in malignancy demonstrate adaptation of the cell genome to its new conditions or environment. These changes are similar *in vivo* to those found *in vitro* in established cell lines (Levan, 1970). From this standpoint, chromosome changes are a consequency of malignant transformation. However, aneuploidy may also be the cause of malignancy: possibly, in special cases (CML), cancer would not occur if certain chromosome abnormalities were not present. What arguments favor one or the other theory? One possible approach is to ascertain some common chromosome patterns in different varieties of tumors and to discuss their meaning.

NEOPLASIAS WITH NORMAL CHROMOSOME COMPLEMENT

The existence of malignant tumors with apparently normal chromosomes shows that the transformation of the normal cell into the malignant cell does not necessitate apparent changes of karyotype. However, this does not signify absence of adaptation of the genome; it means only that our techniques do not permit their detection. It allows wide speculation such as the occurrence of point mutation, small deletion or translocation, derepression of normally repressed genes, or other possible mechanisms. Cytogenetics is useful in providing indirect evidence that all these hypotheses may apply.

NEOPLASIAS WITH SPECIFIC CHROMOSOME ABNORMALITY

In man, there is only one example of a specific abnormality linked with malignant disease: the Ph^1 chromosome in chronic myeloid leukemia (CML). Others markers have been observed in various other forms of neoplasia, but their specificity can not be compared with the Ph^1 chromosome. A large marker has been observed in testicular tumors (Martineau, 1966; Galton et al., 1966), a large submetacentric in some cancers of the cervix uteri (Auersperg et al., 1966, 1967), an acrocentric the size of a B chromosome in ovarian cancer (Lejeune and Berger, 1966), a deleted Gp- or Christchurch chromosome Ch^1 in chronic lymphoid leukemia (Gunz et al., 1962), deleted Eq- or Melbourne chromosome M^1 in malignant lymphomas (Spiers and Baikie, 1966), and a large marker in Waldenström's macroglobulinemia (see Chap. X) which is not, however, a neoplastic disorder. They have been described in only a few patients and their presence has not been confirmed by extensive research.

Is the Ph^1 chromosome to be considered the *cause* of chronic myeloid leukemia? It must be stated here that if the Ph^1 chromosome is characteristic of CML, then CML is not really a characteristic cancer. What kills the patient is the blastic crisis of CML, not the chronic phase of the disease. Aneuploid cell lines are much more frequently observed during the blastic crisis than in the chronic phase. Thus the marker chromosome conceivably places the cell into a "state of chromosome instability" (Blaikie, 1965), which, in turn, favors the action of another factor which gives the disease its characteristic fatal evolution. However, other explanations are possible.

When deletion of one chromosome is present, the possibility exists that one or more recessive genes on the homologous chromosome may express themselves (Elmore et al., 1966). This situation offers opportunity for "deletion-mapping" (McKusick, 1968). Since the specific chromosome abnormality in CML is a deletion, one cannot exclude the expression of recessive genes on the other chromosome of the pair. In fact, Randall et al. (1965) described a familial myeloproliferative disorder "closely simulating childhood myeloid leukemia." Of nine related young children (first or second cousins) having the disorder, at least two recovered completely. No consistent chromosomal aberration was found. A recessive trait was postulated; it is interesting to note that leukocyte alkaline phosphatase was low, as in CML, in four of four affected children and in 18 of 20 asymptomatic relatives.

Normal chromosomes are thus an acceptable finding in CML, since an acquired recessive mutation on *both* chromosomes No. 22 could explain Ph^1 negative CML, and deletion without a mutant gene on the homologous chromosome could explain the Ph^1 chromosome without CML. X-radiation could favor CML because it has both breaking and mutating effects. This hypothesis shows that even in the presence of a specific marker, point mutation cannot be ruled out. In fact, discussion of the subject "chromosomes and cancer" cannot be restricted only to the morphology of chromosomes, but must be extended to their content, i.e. the genes, whenever knowledge permits.

NEOPLASIAS WITH NONSPECIFIC ANEUPLOID CELL LINES

This category of neoplasias is represented in the vast majority of reports concerning chromosomes and cancer. At first glance, findings may appear to be very disappointing since even the same type of tumor may exhibit quite different chromosome patterns in different patients. Claims for "common" pathways of chromosome evolution (De Nava, 1969; Gofman *et al.*, 1970) have not been confirmed, even if, at times, it appears that chromosomes of certain groups have been more involved than others.

However, it is accepted that chromosome evolution in cancer, or in tissue culture *in vitro*, is not just due to chance (Levan, 1967). "Malignancy is a heritable state" (editorial, *Nature*, 1970), at least at the cellular level, and the stability of stemlines in human acute leukemias (Reisman *et al.*, 1964) is an example of the importance of certain stemlines in the propagation of the tumor. Thus, at least as a working hypothesis, it may be postulated that different chromosome changes in the same type of tumor have the same meaning, even if they do not have the same morphology. Some *in vitro* experiments seem relevant in this context. Loss of growth control *in vitro*, very similar to loss of growth control of malignant cells *in vivo*, can be reversed in tissue culture. This reversal is frequently accompanied by chromosome changes in the cell. The changes may be due either to fusion of malignant and nonmalignant cells (Harris et al., 1966) or to the increase or decrease in chromosome number (Pollack et al., 1970; Rabinowitz and Sachs, 1970). For Rabinowitz and Sachs (1970) "the balance between two sorts of chromosomes—those carrying factors for the expression and those with factors for the suppression of transforming genes—determines whether or not the malignant phenotype is expressed." Chromosomes which may carry *genes for repressing malignancy* (Rowley and Bodner, 1970) may be in future identified with the newly available cytogenetic techniques. It is quite likely that chromosomes which carry genes for stimulation of malignancy exist. These *stimulator genes* may be on different chromosomes so that different karyotypes, in fact, have the same meaning and the same phenotypic effect.

Beyond any doubt, more refined techniques, such as recently reported methods for detection of satellite DNA, are necessary to progress in the field of chromosomes and cancer (see Chap. VI). However, studies heretofore published and reported in the present monograph have been useful and may be considered an initial approach to the problem of localization of genes repressing or stimulating malignancy.

REFERENCES

Auersperg, N.; Corey, M. J.; Austin, G.: Chromosomes in cervical lesions. *Lancet,* 1:604, 1966

Auersperg, N.; Corey, M. J.; Worth, A.: Chromosomes in preinvasive lesions of the human uterine cervix. *Cancer Res,* 27:1394, 1967.

Baikie, A. G.: Chromosomes and leukaemia. *Acta Haemat.,* 36:157, 1966.

De Nava, C., and Cortinas, D. E.: Les anomalies chromosomiques au cours des hémopathies malignes et non malignes. *Mongr Ann Génet,* (Paris), l'Expansion ed., 1969, vol. 1.

Elmore, S. M.; Nance, W. E.; McGee, B. J.; Engel deMontmollin, M.; Engel, E.: Pycnodysostosis with a familial chromosome anomaly. *Am J Med, 40:* 273, 1966.

Editorial, *Nature, 231*:488, 1971.

Galton, M.; Benirschke, K.; Baker, M.; Atkin, N. B.: Chromosomes of testicular teratomas. *Cytogenetics,* 5:261, 1966.

Gofman, J. M.; Minkler, J. L.; Tandy, R. K.: A specific common chromosomal pathway for the origin of human malignancy. *U of Calif Lawrence Rad Lab Ref 50356,* 1967.

Gunz, F. W.; Fitzgerald, P. H.; Adams, A.: An abnormal chromosome in chronic lymphocytic leukaemia. *Br Med J, 2*:1097, 1962.

Harris, H.; Miller, O. J.; Klein, G.; Worst, P.; Tachibana, T.: Suppression of malignancy by cell fusion. *Nature, 223*:363, 1969.

Lejeune, J., and Berger, R.: Sur une méthode de recherche d'un variant commun des tumeurs de l'ovaire. *C R Acad Sci* (Paris), *262*:1885, 1966.

Levan, A.: Chromosome abnormalities and carcinogenesis. In A. Lima-de-Faria (Ed.): *Handbook of Molecular Cytology*. Amsterdam-London, North Holland Publishing Company, 1969, p. 717.

Levan, G.: Contributions to the chromosomal characterization of the PTK ratkangaroo cell line. *Hereditas, 64*:85, 1970.

McKusick, V.: *Mendelian Inheritance in Man*, 3rd ed. Baltimore, Johns Hopkins Press, 1971.

Martineau, M.: A similar marker chromosome in testicular tumours. *Lancet, 1*:839, 1966.

Pollack, R.; Wolman, S.; Vogel, A.: Reversion of virus-transformed cell lines: hyperploidy accompanies retention of viral genes. *Nature, 228*:938, 1970.

Rabinowitz, Z., and Sachs, L.: Control of the reversion of properties in transformed cells. *Nature, 225*:136, 1970.

Randall, D. L.; Reiquam, C. W.; Githens, J. H.; Robinson, A: Familial myeloproliferative disease. A new syndrome closely simulating myelogenous leukemia in childhood. *Am J Dis Child, 110*:479, 1965.

Reisman, L. E., Trujillo, J. M.; Thompson, R. I.: Further observations on the role of aneuploidy in acute leukemia. *Cancer Res, 24*:1448, 1964.

Rowley, J. D., and Bodmer, W. F.: Relationship of centromeric heterochromatin to fluorescent banding patterns of metaphase chromosomes in the mouse. *Nature, 231*:503, 1971.

Spiers, A. S. D., and Baikie, A. G.: Cytogenetic studies in the malignant lymphomas. *Lancet, 1*:506, 1966.

Chapter III

CONGENITAL CHROMOSOME ABNORMALITIES AND CANCER

THE LINK BETWEEN a specific congenital chromosome abnormality and a specific type of cancer was demonstrated, in fact, before congenital chromosome syndromes were known. Krivit and Good (1956, 1957) documented increased risk of developing acute leukemia in Down's syndrome, an observation since confirmed in other studies (Stewart et al., 1958; Doll et al., 1962). Conversely, examination of large series of acute leukemia have disclosed that three of 49 patients (Reisman et al., 1964) and two of 65 patients (Kiossouglou et al., 1965) had Down's syndrome. Numerous other case reports illustrating the association of trisomy 21 and leukemia are presented in Table II.

TABLE II
CASES OF LEUKEMIA ASSOCIATED WITH TRISOMY 21

Authors	No. of Cases	46	47,21+	47<	Remarks
Borges et al. (1962)	3				No detailed information
Conen and Erkman (1966)	8	1	6	1	
Reisman et al. (1964)	3	—	1	2	
Sandberg et al. (1966)	2	—	—	2	
Tough et al. (1961)	5	—	5	—	
Brown and Prapp (1962)					
German et al. (1962)					
Haylock and Williams (1963)					
Honda et al. (1963)					
Johnston (1961)					Each report concerns
Kahn and Martin (1967)					1 case. One case
Kiossouglou et al. (1964)					with 46 chromosomes
Lahey et al. (1963)	16	2	4	10	(German et al., 1962)
Lejeune et al. (1963)					concerns a trans-
Mercer et al. (1963)					location mongolism.
Petit et al. (1968)					
Ross and Atkins (1962)					
Thompson et al. (1963)					
Vincent et al. (1963)					
Wahrman et al. (1963)					
Warkany et al. (1963)					
TOTAL		37	3	16	15

Note: Cell lines with more than 47 chromosomes show mainly increases in C and G groups.

For better understanding of the possible link between congenital chromosome abnormalities and cancer, the following questions must be answered: Do all trisomic patients develop the same form of leukemia? Do patients with chromosome abnormalities other than trisomy 21 have leukemia? Do patients with other forms of cancer exhibit more congenital chromosome abnormalities than those of the general population? Are congenital chromosome abnormalities more frequent in cancer families?

LEUKEMIA AND TRISOMY 21

Trisomic children frequently show abnormalities of leucopoiesis which may be confused with leukemia (Ross et al., 1963: "ineffective regulation of granulopoiesis masquerading as congenital leukemia"; Engel et al., 1964: "transient congenital leukemia in infants with mongolism"; Behrman et al., 1966: "abnormal hematopoiesis in mongolism"). However, the type of leukemia occurring in infants with trisomy 21 is not uniform. About 90 percent of children affected with leukemia have the acute lymphoblastic type. This is not the case for children with trisomy 21 and leukemia. Only 38.5 percent of these have acute lymphoblastic leukemia. Trisomy 21 thus seems related to leukemia "as a whole," but not to a specific type of leukemia. This could be understood if trisomy 21 acted more on the regulation of leucopoiesis in general than on any particular hematopoietic cell line. In terms of molecular genetics, this could mean that regulator genes seemed more involved than structural genes.

In fact, no structural genes have been demonstrated with certainty on chromosome 21. It was thought that the chromosome involved in CML (Ph^1) is the same as the one involved in Down's syndrome, a hypothesis which has been recently disproved (O'Riordan et al., 1971). (For further discussion, see Chap. VI.)

There is a wide range of karyotype abnormalities in leukemias associated with trisomy 21 (see Fig. 1) similar to the wide range of abnormalities in non-mongoloid patients with leukemia. The general trend, however, is towards hyperploidy (Table II). Conen and Erkman (1966) suggested that the G group chromosomes might be more often involved than chromosomes of other groups, but this is by no means a general rule; there is no specific pattern of chromosome evolution in leukemias associated with trisomy 21.

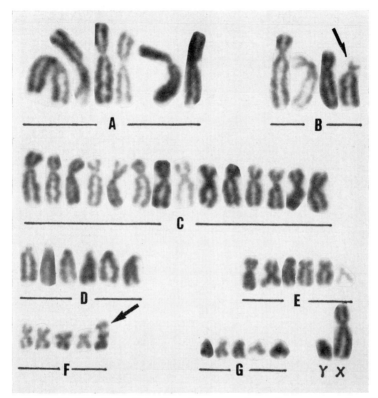

Figure 1. A 48, XY, 21+, F+, Bp− karyotype from the marrow of a child with Down's syndrome and acute myeloblastic leukemia. The arrows point out the Bp− chromosome and the supernumerary F. The Bp− is morphologically identical to the chromosome characteristic of the "cri du chat" syndrome. However, the patient did not exhibit any features of this syndrome and the chromosomes from fibroblasts showed only a standard trisomy 21. Thus, the Bp− chromosome was characteristic of the leukemia cell line in this case, as was the supernumerary F— chromosome with secondary constriction. (Courtesy of Dr. P. Petit, Department of Pathology, University of Brussels.)

LEUKEMIAS IN PATIENTS WITH CONGENITAL CHROMOSOME ABNORMALITIES OTHER THAN TRISOMY 21

Patients with autosomal anomalies other than trisomy 21 may also be affected by leukemia (see Table III). The number of cases is small because many autosomal abnormalities, such as trisomy 13 or 18, do not permit long postnatal survival. Sex chromosome abnormalities are more frequent in the general population, so that comparative data may not be significant.

TABLE III
KARYOTYPES FOUND IN CASES PRESENTING
CONGENITAL CHROMOSOME ABERRATIONS WITH LEUKEMIA

Karyotype	Type of Leukemia	Authors
47, XXY	AL	Kemp et al. (1961); Mammunes et al. (1961) Bouser and Tanzer (1963); Borges et al. (1966)
47, XXY	CML	Tough et al. (1961)
47, XXY	AL	Borges et al. (1966)
, XO/ , XX	CLL	Dumars (1967)
, XO/ , XXX	AL	Lewis et al. (1963)
47, D+	AL	Schade et al. (1962)
45, t (D/D)	CML	Engel et al. (1965)
47, F+	AL	Borges et al. (1966)

CONGENITAL CHROMOSOME ABNORMALITIES AND CANCERS OTHER THAN LEUKEMIA

A relationship between several types of cancer and specific chromosome abnormalities has been suspected for a long time. Patients with Klinefelter's syndrome may have a greater frequency of breast cancer (Jackson et al., 1965), although this has not been definitely proved (Nodel and Koss, 1967); tumors of the gonads and endometrium appear to be more frequent in females with gonadal dysgenesis linked with abnormalities of the sex chromosomes (Teter and Boczkowski, 1967; Dumars, 1967). Sporadic cases of malignant tumors in carriers of a congenital chromosome abnormality have been described (see Fig. 2.). A systematic approach of this problem has been undertaken by Harnden et al. (1969). The chromosomes of 1149 patients with malignant diseases admitted during one year to the Department of Radiotherapy, Western General Hospital in Edinburgh were studied. Eight patients showed a definite chromosome abnormality (see Table IV).

Despite the fact that this number of chromosomal abnormalities (0.69%) was higher than that expected in the general population, the conclusion of the authors was that "this frequency is only slightly higher than that found in normal populations and no special significance is attached to this increase." Still more work is needed and Harnden proposed to study special groups of tumors, such as breast cancers, to gain new insight into this problem.

At this point, it may be concluded that only leukemia seems to be linked with trisomy 21. Patients with Down's syndrome, do not show

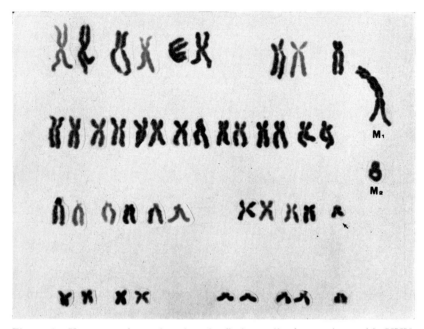

Figure 2. Karyotype from the pleural effusion cell of a patient with XXY-Klinefelter's syndrome and lung carcinoma. Note the large dicentric marker M_1 and ring M_2 probably involving some of the two missing chromosomes of the X-C group and B and E.

TABLE IV

KARYOTYPES AND TYPES OF CANCERS FOUND IN 1149 PATIENTS ATTENDING THE DEPARTMENT OF RADIOTHERAPY OF WESTERN GENERAL HOSPITAL IN EDINBURGH

Karyotype	Type of Tumor
45, t(D/D)	lymphosarcoma
45, t(D/D)	carcinoma, breast
46, inv. C(peq+)	carcinoma, right lung
47, XY, mar +	squamous carcinoma, oral
47, XXX	carcinoma, right breast
46, X inv. (p+g)	lymphosarcoma
46, XX, t(2q—(q+))	papillary carcinoma, right ovary

Note: From Harnden et al., 1969.

a significant increase in other types of cancer (Holland et al., 1962; Miller, 1966).

FAMILIAL AGGREGATION OF CONGENITAL ABNORMALITIES AND CANCER

Families have been described in which several members have exhibited various forms of congenital abnormalities while others have shown a certain form of cancer. Thus in a family described by Baikie et al. (1961), the parents were normal, one child died from acute lymphoblastic leukemia, one from acute myeloblastic leukemia, one from bronchopneumonia, and one was a mosaic of the type XY/XXY. In another family studied by Buckton et al. (1961), the mother was a D/G translocation carrier: she had four children and two were normal, one was 21-trisomic, and one had acute leukemia (AL). Hungerford and Nowell (1961) reported a family in which the father had chronic lymphocytic leukemia, a son had an XXXXY karyotype, and an aunt and a cousin had trisomy 21. Thompson et al. (1963) observed a child with trisomy 21 and acute leukemia. Vereesen et al. (1964) studied a family in which the mother was a mosaic normal/trisomy 21, her son had standard trisomy 21, and her father and two paternal uncles had chronic myeloid leukemia. Other examples are known: five cases of AL and one case of mongolism in three generations (Heath and Maloney, 1965); three siblings—two with mongolism and the third with AL (Conen et al., 1966). A possible tendency to nondisjunction has been reported in the family of a mongoloid child (Kiossouglou et al., 1965). "Prezygotic determinants in acute leukemias" have been suggested by Borges et al. (1967).

It is tempting to suggest that some factors are common to teratogenesis and oncogenesis. If aneuploidy is associated with cancer, genes such as "sticky" in maize (Beadle, 1932) or "$s\ d$" (segregation distorter) in Drosophila, which favor aneuploidy, would also favor in some way the transformation of a normal cell into a malignant cell. Aneuploidy in gametes would be only a special case of this tendency toward aneuploidy. From this point of view, congenital chromosome abnormalities and cancer in the same family would be a different expression of the same fundamental disease. However, this is merely a hypothesis. A systematic study conducted by Holland et al. (1962) failed to show any statistically significant increase of cancer among parents of children with Down's syndrome.

Somewhat different are the examples of congenital chromosome abnormalities and diseases with frequent malignant transformation. A triplo-X mother of a child with Fanconi's anemia showing breaks was reported by Bloom et al. (1966). A mother with a W chromosome marker (see Fig. 15) and Waldenström's macroglobulinemia had a boy with D/D translocation (Lustman et al., 1968); another woman, with a similar marker and with an increased level of gamma-globulins, had a child with trisomy 21 (Elves and Israels, 1963). It has been suggested that auto-immune diseases, especially thyroid auto-immunity, seem to be linked, either in the patient himself, or in his parents, with congenital chromosome abnormalities. Fialkow (1967) suggested that impairment of immunologic mechanisms predispose to certain congenital chromosome disorders, and some types of cancer may be linked. Certainly cytogenetics will be a puissant tool for investigation of these problems. However, no conclusion can be reached at the moment.

Study of possible genetic determinants in cancer further includes twin and family studies.

TWIN STUDIES

For technical reasons, few twin studies have been carried out on hematological disorders. It is known that if a monozygotic (MZ) twin develops leukemia, there is a 20 percent probability that the other twin will also develop leukemia. No such risk for dizygotic (DZ) twins was observed (MacMahon and Levy, 1964). Moreover, from a review of the literature, Pearson et al. (1963) concluded that "it seems likely that the true incidence of concordant leukemia in identical twins is greater than can be explained by chance, but considerably less than could be expected if strong genetic factors were operating." Bone marrow chromosomes of four pairs of identical twins, discordant for chronic myeloid leukemia have been studied. In each pair, the Ph^1 chromosome was found in the leukemic twin, but not in the other (Goh and Swisher, 1965; Dougan et al., 1966; Jacobs et al., 1966; Goh et al., 1967). This clearly demonstrates that both CML and the Ph^1 chromosome were acquired and not inherited. The same observation has been made in Waldenström's macroglobulinemia: only the affected twin showed a marker chromosome in blood lymphocytes (Spengler et al., 1966). A similar observation in DZ twins, one with

AL and the other normal, has been reported by Kiossouglou *et al.* (1964): the twin with AL was the only one to show an abnormal stemline. The chromosomes of one pair of MZ male twins, both with acute leukemia, were studied by Pearson *et al.* (1963): At the time of the study, one child was in relapse and the other in remission. In the first child, an abnormal cell line with 65 chromosomes was observed; in the second, of 130 mitoses studied, one had 64 and three had 65 chromosomes. These results could indicate that a similar cell line was present in both twins, that both patients reacted the same way to the leukemogenic agent, and that chromosome evolution patterns could be genetically predetermined. This finding is in contrast with that found in DZ twins. Sandberg *et al.* (1966) studied chromosomes of DZ twins, both affected with leukemia. One had a stemline with 45 chromosomes and the other one had a stemline with 52 chromosomes. Each twin reacted to the leukemogenic agent as two genetically different individuals, showing different stemline karyotypes, yet displaying the same clinical disease.

From this data it seems possible to conclude that chromosome abnormalities in leukemia, and perhaps in cancer in general, are acquired and not inherited. Chromosome evolution patterns are influenced by the genome of the patient.

FAMILY STUDIES

Family studies in cancer are difficult in many respects. Cancer families seem to exist, but available data do not allow generalization. Review of our present knowledge has been offered in a monograph by Lynch (1967). Indeed, some Mendelian inherited diseases are not infrequently associated with cancer: neurofibromatosis with sarcoma, enchondromatosis with chondrosarcoma, Gardner's syndrome, i.e. familial intestinal polyposis with adenocarcinoma, tuberous sclerosis with glioma, Bloom's syndrome with leukemia, Fanconi's aplastic anemia with leukemia or solid tumors, Bruton's agammaglobulinemia with acute lymphoblastic leukemia, ataxia-telangiectasia with lymphoreticulosarcoma, Chediak-Higashi syndrome with lymphoma, etc.

Twin studies would suggest that in "cancer families," only the affected patient would show a unique chromosomal pattern. This, however, is far from being the case in the few families reported. At times, increase of a specific chromosome anomaly such as breaks, gaps, rings, or other chromosome rearrangements can be observed in non-affected relatives. At other times, more "specific" abnormalities are noticed in apparently healthy relatives. Thus, a peculiar G chromosome with deleted short arms has been reported in many healthy members of a family, while other relatives have presented chronic lymphoid leukemia (Gunz et al., 1962). Weiner (1965) observed a Ph^1 chromosome in a daughter and two of three grandchildren of a patient with Ph^1 positive CML. At the time of the observation, all were healthy. Brown et al. (1967) noticed the presence of a characteristic marker in blood cultures of a patient with Waldenström's macroglobulinemia and in four healthy relatives. The sister of the proband, following encephalitis, showed an increase in serum globulin level with a corresponding increase of 8 percent of cells having the marker. Levin et al. (1967) observed a "Ph^1-like" chromosome in two brothers with polycythemia vera.

The significance of these observations is not clear. At first glance, they appear to be in contradiction to results of twin studies. Nonspecific abnormalities are difficult to interpret, since they can be considered abnormal only on statistical grounds. More "specific" abnormalities, such as the presence of a marker in healthy individuals, might mean, in fact, that this individual will sooner or later develop the disease. Follow-up is then very important in order to assess this point. In the case of a Ph^1 (or "Ph^1-like") chromosome, it has not been possible to rule out the presence of a congenital chromosome abnormality (e.g. a balanced translocation) which has a very different meaning than an acquired chromosome abnormality. In Weiner's family, the leucocyte alkaline phosphatase (LAP), generally lowered in CML, was normal in subjects with the Ph^1-like chromosome. Levin et al. (1967) could not decide the possible meaning of the Ph^1 chromosome observed in two brothers with polycythemia vera.

Furthermore, it is relevant to note that a small deleted Gq- ("Ph^1-like") chromosome has been reported several times in congenital malformation syndromes (Pfeiffer et al., 1962; Dent et al., 1963; Lejeune et al., 1964; Dubois et al., 1965; Goh, 1964; and others).

At present, familial incidence of Ph_1-chromosome is still a subject of discussion and more research is needed before any conclusions can be reached.

REFERENCES

Baikie, A. G.; Buckton, K. E.; Court-Brown, W. M.; Harnden, D. G.: Two cases of leukemia and a case of sex chromosome abnormality in the same sibship. *Lancet, 2*:1003, 1961

Beadle, G. W..: A gene for sticky chromosome in Zea mays. *Z indukt Abstamm Vererblehre, 63*:195, 1932.

Behrman, R. E.; Sigler, A. T.; Ratchefsky, A. S.: Abnormal hematopoiesis in 2 or 3 siblings with mongolism. *J Pediat, 68*:569, 1966.

Bloom, G. E.; Warner, S.; Gerald, P. S.; Diamond, L. K.: Chromosome abnormalities in constitutional aplastic anemia. *N Engl J Med, 274*:8, 1966.

Borges, W. H.; Wald, N.; Kim, J.: Non-specificity of chromosomal abnormalities in human leukemia. *Clin Res, 10*:211, 1962.

Borges, W. H.; Nicklas, J. W.; Hamm, C. W.: Prezygotic determinants in childhood leukemias. *J Pediatr, 68*:837, 1966.

Borges, W. H.; Nicklas, J. W.; Hamm, C. W.: Prezygotic determinants in acute leukemia. *J Pediatr, 70*:180, 1967.

Bousser, J., and Tanzer, J.: Syndrome de Klinefelter et leucémie aigüe. *Nouv Rev Franc Hemat, 3*:194, 1963.

Brown, C. D., and Propp, S.: Trisomy for chromosome 21 and tetraploid metaphases in congenital acute granulocytic leukemia in mongolism.. *Hum Chr Newsl, 7*:12, 1962.

Brown, A. K.; Elves, M. W.; Gunson, H. H.; Pell-Ilderton, R.: Waldenström's macroglobulinemia: A family study. *Acta Haemat, 38*:184, 1967.

Buckton, K. E.; Harnden, D. G.; Baikie, A. G.; Woody, G. E.: Mongolism and leukemia in the same sibship. *Lancet, 1*:171, 1961.

Conen, P.E., and Erkman, B.: Combined mongolism and leukemia. *Am J Dis Child, 112*:429, 1966.

Conen, P. E.; Erkman, B.; Laski, B.: Chromosome studies on a radiographer and her family. *Arch Intern Med, 117*:125, 1966.

Dent, T.; Edwards, J. H.; Delhanty, J. D. A.: A partial mongol. *Lancet, 2*: 484, 1963.

Doll, R.; Holland, W. W.; Carter, C. O.: The mortality from leukemia and other cancers among patients with Down's syndrome (mongols) and their parents. *Br J Cancer, 16*:177, 1962.

Dougan, L.; Scott, I. D.; Woodliff, H. J.: A pair of twins, one of whom has chronic granulocytic leukemia. *J Med Genet, 3*:217, 1966.

Dumars, K. W.: Cancer, chromosomes and congenital abnormalities. *Cancer, 20*:1006, 1967.

Elves, M. W., and Israels, M. L. G.: Chromosomes and serum proteins: a linked abnormality. *Br Med J, 1*:1024, 1963.

Engel, E.; Hammond, D. D.; Eitzman, D. V.; Pearson, H.; Krivit, W.: Transient congenital leukemia in 7 infants with mongolism. *J Pediatr, 65*: 303, 1964.

Engel, E.; McGee, R. J.; Hartmann, R. C.; Engel-Demontmollin, M.: Two leukemic peripheral blood stemlines during acute transformation of chronic myelogenous leukemia in a D/D translocation carrier. *Cytogenetics, 4*:157, 1965.

Fialkow, P. J.: Immunologic oncogenesis. *Blood, 30*:388, 1967.

German, J. L.; Demayo, A. P.; Bearn, A. G.: Inheritance of an abnormal chromosome in Down's syndrome (mongolism) with leukemia. *Am J Hum Genet, 14*:31, 1962

Goh, K. O.: Abnormal chromosome signals myelocytic leukemia. *JAMA, 188* (suppl):37, 1964.

Goh, K. O., and Swisher, S. N.: Identical twins and chronic myelocytic leukemia. Chromosomal studies of a patient with chronic myelocytic leukemia and his normal identical twin. *Arch Intern Med, 115*:475, 1965.

Goh, K. O.; Swisher, S. N.; Herman, E. C.: Chronic myelocytic leukemia and identical twins. *Arch Intern Med, 120*:214, 1967.

Gunz, F. W.; Fitzgerald, P. H.; Adams, A.: An abnormal chromosome in chronic lymphatic leukemia. *Br Med J, 2*:1097, 1962.

Harnden, D. G.; Langlands, A. D.; McBeath, S.; O'Riordan, M.; Faed, M. J. X.: The frequency of institutional chromosome abnormalities in patients with malignant diseases. *Europ J Cancer, 5*:605, 1969.

Haylock, J., and Williams, A.: Abnormal chromosome patterns in a children's hospital. *Med J Austr, 50*:184, 1963.

Heath, C. W., and Moloney, W. C.: Familial leukemia. Five cases of acute leukemia in three generations. *N Engl J Med, 272*:882, 1965.

Holland, W. W.; Doll, R.; Carter, C. O.: The mortality from leukemia and other cancers among patients with Down's syndrome (mongols) and among their parents. *Br J Cancer, 16*:177, 1962.

Honda, F.; Punnett, H. P.; Charney, E.; Miller, G.; Thiede, H. A.: Serial cytogenetic and hematologic studies on a mongol with trisomy 21 and acute congenital leukemia. *J Pediatr, 65*:880, 1964.

Hungerford, D. A., and Nowell, P. C.: Chromosome studies in human leukemia. I. Acute leukemia in children. *J Nat Cancer Inst, 27*:983, 1961.

Jackson, A. W.; Muldal, S.; Ockey, C. H.; O'Connor, P. J.: Carcinoma of male breast in association with the Klinefelter syndrome. *Br Med J, 1*:223, 1965.

Jacobs, E. M.; Luce, J. K.; Cailleau, R.: Chromosome abnormalities in human cancer. Report of a patient with chronic granulocytic leukemia and his non-leukemic monozygotic twin. *Cancer, 19*:869, 1966.

Johnston, A. W.: The chromosomes in a child with mongolism and acute leukemia. *N Engl J Med, 264*:591, 1961.

Kahn, M. H., and Martin, H.: G21 trisomy in a case of acute myeloblastic leukemia. *Acta Haemat, 38*:142, 1967.

Kemp, N. H.; Stafford, J. L.; Tanner, R. K.: Acute leukemia and Klinefelter's syndrome. *Lancet, 2*:434, 1961.

Kiossouglou, K. A.; Rosenbaum, E. H.; Mitus, W. S.; Dameshek, W.: Multiple chromosomal aberrations in a patient with acute granulocytic leukemia associated with Down's syndrome and twinning. Study of a family with a possible tendency to non-disjunction. *Blood, 24*:134, 1964.

Krivit, W., and Good, R. A.: The simultaneous occurrence of leukemia and mongolism. *Am J Dis Child, 91*:218, 1956.

Krivit, W., and Good, R. A.: Simultaneous occurrence of mongolism and leukemia. *Am J Dis Child, 94*:289, 1957.

Krogh Jensen, M.: *Chromosome Studies in Acute Leukemia.* Munksgaard, Copenhagen, 1969.

Lahey, M. E.; Beier, F. R.; Wilson, J. F.: Leukemia in Down's syndrome. *J Pediatr, 63*:189, 1963.

Lejeune, J.; Berger, R.; Haines, M. Lafourcade, J.; Vialatte, J.; Satge, P.; Turpin, R.: Constitution d'un clone à 54 chromosomes au cours d'une leucoblastose chez une enfant mongolienne. *C R Acad Sci, 256*:1195, 1963.

Lejeune, J.; Berger, R.; Rethore, M. O.; Archambault, L.; Jérôme, H.; Thieffry, S.; Aicardi, J.; Broyer, M.; Lafourcade, J.; Cruveiller, J.; Turpin, R.: Monosomie partielle pour un petit acrocentrique. *C R Acad Sci, 259*:4187, 1964.

Levin, W. C.; Houston, E. W.; Ritzman, S. E.: Polycythemia vera with Ph^1 chromosomes in two brothers. *Blood, 30*:503, 1967.

Lewis, F. J. W.; Poulding, R. H.; Eastham, R. D.: Acute leukemia in an XO/XXX mosaic. *Lancet, 2*:306, 1963.

Lustman, F.; Stoffels-DeSaint Georges, A.; Ardichvili, D.; Koulischer, L.; Demol, H.: La macroglobulinémie de Waldenström. *Acta Clin Belg, 23*: 67, 1968.

Lynch, H. T.: *Hereditary Factors in Carcinoma.* New York-Heidelberg, Springer-Verlag, 1967, vol. 1.

MacMahon, B., and Levy, M. A.: Prenatal origin of childhood leukemia: evidence from twins. *N Engl J Med, 270*:1082, 1964.

Mamunes, P.; Lapidus, P. H.; Abbot, J. A.; Roath, S.: Acute leukemia and Klinefelter's syndrome. *Lancet, 2*:26, 1961.

Mercer, R. D.; Keller, M. K.; Lonsdale, D.: An extra abnormal chromosome in a child with mongolism and acute myeloblastic leukemia. Report of a case. *Cleveland Clin Quart, 30*:215, 1963.

Miller, R. W.: Relation between cancer and congenital defects in man. *N Engl J Med, 275*:87, 1966.

Nodel, M., and Koss, L. G.: Klinefelter's syndrome and male breast cancer. *Lancet, 2*:366, 1967.

O'Riordan, M. L.; Robinson, J. A.; Buckton, K. E.; Evans, H. J.: Distinguishing between the chromosomes involved in Down's syndrome (trisomy 21) and chronic myeloid leukemia (Ph^1) by fluorescence. *Nature, 230*:167, 1971.

Pearson, H. A.; Grello, F. W.; Cone, T. E.: Leukemia in identical twins. *N Engl J Med, 268*:1151, 1963.

Petit, P.; Maurus, R.; Richard, J.; Koulischer, L.: Chromosome du cri du chat chez un trisomique 21 leucémique. *Ann Génét, 11*:125, 1968.

Pfeiffer, R. A.; Schellong, G.; Kosenow, W.: Chromosomenanomalien in den Blutzellen eines Kindes mit multiplen Abartungen. *Klin Wschr, 40*:1058, 1962.

Reisman, L. E.; Trujillo, M. M.; Thompson, R. I.: Further observations on the role of aneuploidy in acute leukemia. *Cancer Res, 24*:1448, 1964.

Ross, J. D., and Atkins, L.: Chromosomal anomaly in a mongol with leukemia. *Lancet, 2*:612, 1962.

Sandberg, A. A.; Cortner, J.; Takagi, N.; Moghadam, M. A.; Crosswhite, L. H.: Differences in chromosome constitution of twins with acute leukemia. *N Engl J Med, 275*:809, 1966.

Sandberg, A. A.; Ishihara, T.; Miwa, T.; Hauschka, T. S:. The *in vivo* chromosome constitution of marrow from 34 human leukemias and 60 nonleukemic controls. *Cancer Res, 21*:678, 1961.

Schade, H.; Schoeller, L.; Schultze, K. W.: D-Trisomie (Patau-Syndrom) mit kongenitaler myeloischer Leukämie. *Med Welt, 50*:2690, 1962.

Spengler, G. A.; Siebner, H.; Riva, G.: Chromosomal abnormalities in macroglobulinemia Waldenström: discordant findings in uniovular twins. *Acta Med Scand, 445*(suppl):132, 1966.

Stewart, A. M.; Webb, J.; Hewitt D.: A survey of childhood malignancies. *Br Med J, 2*:1495, 1958.

Teter, J., and Boczkowski, K.: Occurrence of tumors in dysgenetic gonads. *Cancer, 20*:1301, 1967.

Thompson, M. W.; Bell, R. E.; Little, A. S.: Familial 21-trisomic mongolism coexistent with leukemia. *Can Med Assoc J, 88*:893, 1963.

Tough, I. M.; Court-Brown, W. M.; Baikie, A. G.; Buckton, K. E.; Harnden, D. G.; Jacobs, P. A.; King, M. J.; McBride, J. A.: Cytogenetic studies in chronic myeloid leukemia and acute leukemia associated with mongolism. *Lancet, 1*:411, 1961.

Vereesen, H.; Berghe, van den, H.; Creemers, J.: Mosaic trisomy in phenotypically normal mother of mongol. *Lancet, 1*:526, 1964.

Vincent, P. C.; Sinha, S.; Neate, R.; denDuld, G.; Turner, B.: Chromosome abnormalities in a mongol with acute myeloid leukemia. *Lancet, 1*:1328, 1963.

Wahrman, J.: Mongolism. *Harefuah* (Hebr), *64*:260, 1963.

Warkany, J.; Schuber, W. K.; Thompson, J. N.: Chromosome analyses in mongolism (Langdon Down syndrome) associated with leukemia. *New Eng J Med, 268*:1, 1963.

Weiner, L.: A family with high incidence leukemia and unique Ph[1] chromosome findings. *Blood, 26*:871, 1965 (abstract).

Chapter IV

VIRUSES AND CHEMICAL CLASTOGENS

Since 1961, studies of the role played by viruses in somatic mutation have yielded large amounts of data. The main theme of most publications has been *in vitro* transformation of cells of experimental animals and eventually of cells of established cell lines, including studies of the effects of viruses on chromosomes. It was observed that certain viruses are capable of producing at least three types of chromosome changes: (a) single breaks, (b) pulverization, and (c) fusion and spindle abnormalities (Nichols, 1966). Interest in the role of viruses in cancerogenesis arose mainly following Rous' demonstration that a cancer in chickens could be provoked by a special type of virus. The most typical chromosomal aberration caused by a virus was found to be a break, being first reported in herpes simplex virus-infected cells of Chinese hamster *in vitro* (Hampar and Ellison, 1961). Then followed a number of reports of other types of viruses causing breaks such as Rous sarcoma virus (Nichols, 1963), measles virus (Fjelde and Holterman, 1962), herpes zoster virus (Benyesh-Melnick *et al.*, 1964), Sendai virus (Saksela *et al.*, 1965), chicken pox and mumps (Aula, 1963, 1965), and others. Positive observations were also made with yellow fever vaccine (Harden, 1964), aseptic meningitis (Makino *et al.*, 1965), infectious hepatitis (El-Alfi *et al.*, 1965), polyoma virus (Vogt and Dulbecco, 1963), and foot-and-mouth disease virus (Stenkvist *et al.*, 1965).

Several mechanisms by which virus could produce a chromosome break were postulated including direct virus combination with the chromatid, interference with DNA synthesis either by inhibiting cellular enzymes or competing for nucleic acid building blocks—or by producing an effect on a cellular organelle, such as the lysosome, that produces its action through release of DNA-ase (Nichols, 1966). None of these hypotheses seems to be acceptable with exclusion of other possibilities. Recently there are further indications that specific heterochromatic areas might be predilection sites of virus association with chromosomes (Maio, 1971), a fact already appreciated in the early sixties by Moorhead and Saksela (1963) and Hsu (1963). The meaning of viral association with a heterochromatic region is still unknown, but the effect could more likely be explained as the addition to the cell of the viral genome than by hypothesizing alteration of the cell's own genome (see Chap. VI).

At present there is no human malignant tumor in which the virus has been proven to be the causative agent. There are, however, strong indications that the viral theory of origin of human cancers might be valid. Further investigation of *Burkitt's lymphoma* may possibly substantiate the proof. In this tumor, virus or virus-like particles (Epstein-Barr virus) have been commonly demonstrated in affected tissues and in cultures. Limited geographical distribution of the tumor to wet tropical areas also indicates the possibility of its induction by arthropod-borne virus.

We have reviewed five reports on chromosome analysis of Burkitt's tumor from *direct* preparations. First, Jacobs et al. (1963) analyzed nine cases by the direct method. Four had apparently normal karyotypes and five displayed long acrocentric markers approximately the size of the long arms of a No. 2 chromosome. Interestingly, a similar marker was reported by Stewart et al. (1965). Chu et al. (1966) examined ascitic fluid from a patient with a Burkitt-like tumor and, in 90 percent of mitoses, found a long submetacentric marker similar to that later described by Gripenberg et al. (1969) and was also observed in a large proportion of cells in cases reported by Jacobs et al. (1963) and in a single case by Clifford et al. (1968).

None of the authors observed any aberrant mitoses either in peripheral blood or in bone marrow. There were several cases with cells of normal diploid complement. Aberrant mitoses in other cases were mainly pseudodiploid or near diploid; the only exception was one case with 80 percent tetraploid cells reported by Clifford and colleagues (1968).

It would be premature to attribute much significance to similar large markers so frequently observed. It remains to be seen whether there will be more consistent reports in the future, especially regarding diploid or near-diploid karyotypes and the frequent aberration of

A group chromosomes. Surprisingly, the most typical virus-induced type of unstable aberrations—the break—was not a remarkable feature of reported karyotypes.

A large number of agents have been suspected of teratogenic or cancerogenic effect. Many of these physical, chemical, and biological agents are known to break chromosomes. Shaw (1970) introduced the term *clastogen* to emphasize their effect on chromosomes. The chemical clastogens are numerous and their action is not always regarded as being responsible for development of either cancer or congenital malformations.

One well-recognized chemical clastogen, benzene, attracted the attention of several investigators. Tough and Court Brown (1965) showed that in 20 probands with a history of long-term (1 to 20 years) exposure to benzene vapors, the number of chromosome breaks was significantly increased. An analogous finding was reported by Forni and Moreo (1967) who performed a series of chromosome analysis in a patient with 22 years of benzene exposure. This patient developed severe anemia, at which stage 47 chromosomes were found in his aplastic bone marrow. In eight consecutive analyses of this patient, the chaotic development of his karyotype was seen. The search for a Ph^1-like chromosome revealed this marker in only two of 150 mitoses. Similarly, no consistent finding was reported by Pollini and Colombi (1964) and Hartwick et al. (1969) in two other patients with benzene-induced leukemia. All authors, however, observed a significantly higher number of breaks, gaps, and even minute elements. In this chapter we should make note of the chromosomal examination of a patient with acute leukemia following repeated use of LSD (lysergic acid diethylamide), mescaline, marihuana, and amphetamines (Grossbard et al., 1968). This patient also had hereditary spherocytosis. All 35 mitoses examined showed a Ph^1-like element in 24-hour blood culture without PHA stimulation.

It should be emphasized that production of chromosome breaks is only the first step of a two-step process which results in structural rearrangement of chromosomes (Shaw, 1970). Thus, if these new stable and structurally (hence genetically) different chromosomes arise, their effect might theoretically lead to establishment of a functionally aberrant malignant cell line.

REFERENCES

Aula, P.: Chromosome breaks in leukocytes of chickenpox patients. Preliminary communication. *Hereditas, 49*:451, 1963.
Aula, P.: Virus-associated chromosome breakage. A cytogenetic study of chickenpox, measles and mumps patients and of cell cultures infected with measles virus. *Ann Acad Sci Fenn*, Ser. A. IV *Biologica*, 89, 1965.
Benyesh-Melnick, M.; Stich, H. F.; Rapp, F.; and Hsu, T. C.: Viruses and mammalian chromosomes. III. Effect of herpes zoster virus on human embryonal lung cultures. *Proc Exp Biol Med, 117*:546, 1964.
Chu, E. W.; Whang, J. J. K.; Rabson, A. S.: Cytogenetic studies of lymphoma cells from an American patient with a tumor similar to Burkitt's tumors in African children. *J Nat Cancer Inst, 37*:885, 1966.
Clifford, P.; Gripenberg, U.; Klein, E.; Fenyö, E. M.; Manolov, G.: Treatment of Burkitt's lymphoma. *Lancet 2*:517, 1968.
El-Alfi, O. S.; Smith, P. M.; Biesele, J. J.: Chromosomal breaks in human leukocyte cultures induced by an agent in the plasma of infectious hepatitis patients. *Hereditas, 52*:285, 1965.
Fjelde, A., and Holtermann, O. A.: Chromosome studies in HEp-2 tissue culture cell line during infection with measles virus. *Life Sci. 12*:683, 1962.
Forni, A., and Moreo, L.: Cytogenetic studies in a case of benzene leukaemia. *Europ J Cancer, 3*:251, 1967.
Gripenberg, U.; Levan, A.; Clifford, P.: Chromosomes in Burkitt lymphomas. I. Serial studies in a case with bilateral tumors showing different chromosomal stemlines. *Int J Cancer, 4*:334, 1969.
Grossbard, L.; Rosen, D.; McGilvray, E. et al.: Acute leukemia with Ph^1-like chromosome in an LSD user. *JAMA, 205*:791, 1968.
Hampar, B., and Ellison, S. A.: Chromosomal aberrations induced by an animal virus. *Nature, 192*:145, 1961.
Harden, D. G.: Cytogenetic studies on patients with virus infections and subjects vaccinated against yellow fever. *Am J Hum Genet, 16*:204, 1964.
Hartwich, G.; Schwanitz, G.; Becker, J.: Chromosomenaberrationen bei einer Benzol-Leukämie. *Deutsch Med Wschr, 94*:1228, 1969.
Hsu, T. C.: Longitudinal differentiation of chromosomes and the possibility on interstitial centromeres. *Exp Cell Res*, suppl., *9*:73, 1963.
Jacobs, P. A.; Tough, I. M.; Wright, D. H.: Cytogenetic studies in Burkitt's lymphoma. *Lancet, 2*:1144, 1963.
Maio, J. J.: DNA strand reassociation and polyribonucleotide binding in the African green monkey, Cercopithecus aethiops. *J Mol Biol, 56*:579, 1971.
Makino, S.; Yamada, K.; Kajii, T.; Chromosome aberrations in leukocytes of patients with aseptic meningitis. *Chromosoma, 16*:372, 1965.
Moorhead, P. S., and Saksela, E.: Non-random chromosomal aberrations in SV_{40} transformed human cells. *J Cell Comp Physiol, 62*:57, 1963.
Nichols, W. W.: Relationship of viruses, chromosomes, and carcinogenesis. *Hereditas, 50*:53, 1963.

Pollini, G., and Colombi, R.: Il danno cromosomico dei linfociti nell' emopatia benzenica. *Med d Lavorno*, *55*:641, 1964.

Saksela, E.; Aula, P.; Cantell, K.: Chromosomal damage of human cells induced by Sendai virus. *Ann Med Exp Biol Fenn*, *43*:132, 1965.

Shaw, M. W.: Human chromosome damage by chemical agents. *Ann Rev Med*, *21*:409, 1970.

Stenkvist, B.; Philipson, D. L.; Pontén, J.: Morphological transformation of calf kidney cells induced by foot-and-mouth disease virus. *Exp Cell Res*, *39*:170, 1965.

Stewart, S. E.; Lovelace, E.; Whang, J. J.; Ngu, V. A.: Burkitt tumor: tissue culture, cytogenetic and virus studies. *J Nat Cancer Inst*, *34*:319, 1965.

Tough, I. M., and Court-Brown, W. M.: Chromosome aberrations and exposure to ambient benzene. *Lancet*, *1*:684, 1965.

Vogt, M., and Dulbecco, R.: Steps in the neoplastic transformation of hamster embryo cells by polyoma virus. *Proc Nat Acad Sci*, *49*:171, 1963.

Chapter V

ANALOGY OF KARYOTYPE EVOLUTION IN CANCER AND IN MAMMALIAN SPECIATION

As seen in Chapter I, in some cases development of chromosome changes in spontaneous as well as in induced tumors can be followed accurately by analysis at regular intervals. Karyotype instability is observed at times, the cells becoming more and more aneuploid until new stability is reached. Steps intermediate between the original and the observed karyotype can be detected or reconstructed. These changes are known as karyotype evolution or clonal evolution of the karyotype, the malignant cells forming a clone.

It has already been pointed out that to describe this evolution, essentially the same terms and the same concepts as those used in evolution of species are employed. For example, Lejeune (1965) wrote that even if sometimes "the succession of the intermediate stages cannot in practice be observed," it is, however, "tempting to reconstruct the history of the clone by assuming that the least aneuploid cells, the nearest to normal, represent the relic, the fossils of this evolution." To reconstruct this evolution is an "approach which scarcely differs from the usual practice of paleontologists."

It was therefore tempting to make comparison between clonal evolution in malignancy and chromosome evolution linked with speciation (Koulischer, 1968). Despite the great difference existing between malignant cells, which reproduce through mitosis, and species, which reproduce through meiosis, some common patterns are observed. The purpose of the present chapter is to outline these similarities. Only mammalian tumors and mammalian speciation will be considered and only chromosome changes linked with speciation and not intraspecific polymorphism will be discussed.

CHROMOSOMES AND SPECIATION IN MAMMALS

Speciation may or may not be linked with chromosome changes. Different models of speciation will be considered in this section.

Speciation With No Apparent Change of Karyotype

This kind of speciation is not unusual. Such phenotypically different species as the cow, buffalo, European wisent, and an African antelope, the common duiker, have the same karotype with $2n = 60$ chromosomes, 58 autosomes being telocentric, and the X chromosomes metacentric (Koulischer *et al.*, 1967). All Felidae of the Northern hemisphere have $2n = 38$ chromosomes with an identical karyotype (Wurster, 1969); all Camelidae including the llamas and guanacos exhibit the same chromosome map (Taylor *et al.*, 1968) which is also true for bears of the Northern hemisphere, for many species of Canidae having the same karyotype as the domestic dog, etc. (Wurster, 1969). Indeed, as classifications cannot and should not be based upon one parameter only (Wurster, 1969), this analogy of karyotypes is not sufficient to affirm that we deal with races instead of species. However, it does mean that speciation may occur without any change of karyotype detectable by our present means of observation.

Robertsonian Fusion of Chromosomes

The interest in evolution by centromeric fusion of acro-telocentric chromosomes was first assessed by Robertson (1916). This mode of speciation is widely employed in many mammalian families and sometimes exclusively, such as in the Bovoidea (Wurster and Benirschke, 1968). There is a general trend for recent species to show a total decrease in chromosome number, a decrease in the number of acro- or telocentrics, and an increase in the number of meta- or submetacentrics. The *nombre fondamental* or NF, that is, the total number of chromosome *arms* in a given species, generally remains constant. This allows easy demonstration of parental relationship between species of the same family (Matthey, 1949). Chromosome fusion linked with speciation is found in so many families that no particular one will be cited except for the Bovoidea. In this family, species with $2n = 60$ chromosomes, all telocentric, can be observed in the goat, together with species with $2n = 30$ chromosomes, al meta- or submetacentric as for an African antelope, the blackbuck (Wurster *et al.*, 1968). The NF is the same: speciation can be explained simply by fusion of telocentrics.

Decrease in Chromosome Number, Without Rearrangements

The Robertsonian fusion of telo- or acrocentric chromosomes is a typical example of a mechanism causing decrease in the diploid number due to centromeric fusion. In some families, however, the decrease in diploid number has the appearance of a loss of chromosomes without visible rearrangements. This could be the case in the family Equidae. The Przewalski horse has $2n = 66$ chromosomes and the domestic horse $2n = 64$ chromosomes. Although other explanations have been proposed (Benirschke et al., 1965), the only difference between the karyotypes of the two species is the apparent loss of telocentric chromosomes in the domestic horse. A similar loss is reported in the South American bush dog (Speothos venaticus) whose karyotype with $2n = 74$ chromosomes is identical to that of the domestic dog except that four telocentrics appear to be missing (Wurster, 1969).

Increase in Chromosome Number

It was recognized that Robertsonian fusion is a rearrangement invariably resulting in a decrease in chromosome number and linked with speciation. Thus, for a long time, it was thought that chromosome increase was not a factor in speciation (the "fission" theory not being widely accepted). However, at least in the Cercopithecae, Chiarelli (1968) demonstrated beyond any doubt that the most recent species were those with the highest chromosome numbers. To explain this unusual observation, he suggested a mechanism of duplication of certain chromosomes of the primitive set with a corresponding increase of DNA content.

The Same Diploid Number With Chromosome Structural Rearrangements

In some families, different species may show the same chromosome numbers; only the karyotypes show significant differences in their morphology. Apparently, rearrangements occurred in the original karyotype, leading to speciation without affecting chromosome number. For example, the chimpanzee, gorilla, and orangutan have $2n =$

48 chromosomes, but different karyotypes. In the family Bovoidea, the blackbuck and sitatunga, two African antelopes, also have an identical chromosome number, 2n = 30, but different karyotypes (Wurster and Benirschke, 1968). Other examples are the Felidae of the Southern hemisphere with 2n = 36 chromosomes (Wurster, 1969), and two species of European hedgehogs, Erinaceus europaeus and Erinaceus roumanicus (Gropp, 1969).

CHROMOSOME CLONAL EVOLUTION IN CANCER CELLS

Chromosome patterns observed in clonal evolution of cancer cells shall be described in this paragraph in the same order as those described for mammalian speciation. Some relevant observations concerning chromosomes of cells in tissue culture shall be considered in this monograph, which is concerned mainly with fresh tumors, since the formation of a permanent tissue culture cell line may be looked upon as a genetic process of adaptation comparable to the change of a normal cell into a cancer cell (Levan, 1970).

Tumors With No Apparent Change in the Original Karyotype

Many human or animal tumors show no apparent changes of karyotype. In human acute leukemias, 50 percent of reported cases have a normal karyotype. All cases of chronic lymphoid leukemia show normal chromosomes, as well as many tumors of the lymphoreticular system and some solid tumors. This demonstrates that chromosome changes are not necessarily involved in transformation of a normal cell into a cancer cell. Similarly, speciation may occur in mammals with very different phenotypes, but with the same karyotype.

Robertsonian Fusion of Acrocentrics

As in mammalian speciation, this particular mechanism has been observed in tumors or tissue cultures *in vitro*. Robertsonian fusions are most frequently detected in species with many acro- or telocentrics, such as cattle or mice. In tumors, Robertsonian fusion seems to be specific in several cases of canine venereal sarcoma (Makino, 1963; Weber et al., 1965; Barski and Cornefert-Jensen, 1966), where reduction in the total number of chromosomes and an increase in metacentrics were consistently observed. The karyotype of the dog has $2n = 78$ chromosomes, 76 being telocentric with the X chromosomes being metacentric. In cells of canine venereal sarcoma, the diploid number is 59, with 42 to 44 acrocentrics and 15 to 17 metacentrics, the NF— (total number of arms) being almost the same as in normal somatic cells. In bovine leukemia, fusion of telocentric chromosomes was also observed (Gustavson and Rockborn, 1964). In human tumors, the frequent increase in the number of C group chromosomes, together with a decrease in the number of D and G group chromosomes, led to the suggestion that this results from fusion and acrocentrics of the D and G groups (Levan, 1967; Muldal et al., 1971).

Apparent Decrease of Chromosome Numbers, Without Rearrangements

In mammalian speciation, this mechanism is not frequently observed. There are, however, some types of human leukemia (erythroleukemia, acute myeloblastic leukemia) in which it is not uncommon to note karyotypes with missing chromosomes. This happens also in some established mammalian cell lines *in vitro*. However, it is known from cytogenetic study of infants with congenital malformations that deletion syndromes are not as frequent as syndromes with supernumerary chromosomes. Tumors, mammalian speciation, and congenital malformations demonstrate that loss of chromosome material is not tolerated as favorably as a gain of chromosomal material.

Apparent Increase in Chromosome Numbers

This mechanism represents the most common finding in tumors and, as mentioned earlier, it is exceptional in mammalian speciation. In fact, most tumors and *in vitro* established cell cultures show one to ten or more supernumerary chromosomes. The increase may be due to polyploidization: in that case, the number of chromosomes is a multiple of n (i.e. the haploid set of chromosomes of the species considered). Polyploidy is so regularly observed in malignant cells (see Fig. 3) that it could be an expression of the better adaptation of the karyotype to environmental factors. In the case of a loss of chromo-

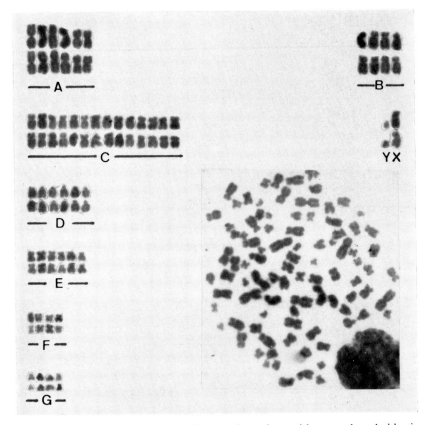

Figure 3. Strict tetraploidy (4 n) in a male patient with acute lymphoblastic leukemia. All mitotic cells from the bone marrow were tetraploid.

somes, which may prove to be fatal to the cell, hypoploidy might be corrected by hypopolyploidy following a two step mechanism: first aneupolidization, then polploidization. Polyploidy is not a mechanism employed in mammalian speciation, as shown also by DNA studies. In fact, all mammals have almost the same nuclear RNA content (Atkin et al., 1965). Polyploidization has occurred in earlier times and may explain evolution from primitive to higher forms of life (Ohno, 1970).

By observing an increase in chromosome numbers in a case of acute leukemia, Lejeune (1965) proposed his laws of clonal evolution mentioned earlier. It is interesting to note that Lejeune for clonal evolution and Chiarelli for mammalian speciation independently proposed the same hypothesis for an increase in chromosome number: *acquisition and duplication of supernumerary autosomes.*

Pseudodiploidy

This kind of rearrangement explains speciation with the same number of chromosomes but with structural chromosome rearrangements. The best known example is the Philadelphia chromosome (Ph^1) in human chronic myeloid leukemia. The number of chromosomes is normal, $2n = 46$, but there is deletion of a G group chromosome. Many other cancers are known to manifest pseudodiploidy. In special cases, it seems that the normal number of chromosomes gives certain advantage to the cell, even if there is inconsistency of karyotype. Thus, Hughes (1968) observed the same normal diploid number of chromosomes in cultured BHK cells after 451 and 480 cell generations, but karyotypes were different from cell to cell so there was no real stemline. Hughes then proposed a "pseudo-stemline concept" in which there was more cooperation than competition among cells.

POSSIBLE MEANING OF THE ANALOGY

Somatic cells reproducing through mitosis could provide models of chromosome evolution in a shorter period of time, compared to chromosome evolution observed in mammalian speciation. From the chromosomal point of view, any drastic karyotype change in a cell with a generation time of 14 to 20 hours can be transmitted directly to the daughter cells and eventually result in a new stemline. Thus, in one generation time of a somatic cell, changes can occur which would need several hundreds, if not thousands of generations in mammalian speciation since mammals must go through repeated meioses to transmit and stabilize any karyotype modification.

Despite those fundamental differences, chromosome evolution in tumors and mammalian speciation shows common morphologic patterns. There are some differences: hyperploidy is more frequent in cellular clonal evolution, Robertsonian fusion is employed more in mammalian speciation, and polyploidy, so commonly observed in cancer cells, is not a factor of speciation in mammals. The possible meaning of the analogy is still open to discussion, but is certainly striking and worthy of consideration.

An attractive hypothesis is that the appearance of a cancerous cell in an organism could be the equivalent of the emergence of a new species inside a mammalian family. Each living species within a family could be considered analogous to a stemline of tissue culture *in vitro*. Thus, the understanding of chromosome mechanisms involved in evolution and speciation of mammals could lead to better understanding of chromosome mechanisms involved in cancer, or vice versa. Some very successful families, such as the Bovoidea, with 130 species, could also provide many more models of chromosome evolution than the study of any tumor or tissue culture in which 130 different stemlines have been never reported. It is possible that the study of chromosomes in mammalian speciation may provide supplementary information and facilitate better understanding of chromosome involvement in cancer, due to the fact that phenotypic differences are more easy to observe in species than in cancer cells.

REFERENCES

Atkin, N. B.; Mattinson, G.; Becak, W.; Ohno, S.: The comparative DNA content of 19 species of placental mammals, reptiles and birds. *Chromosoma, 17*:1, 1965.

Barski, G., and Cornefert-Jensen, F.: Cytogenetic study of Sticker venereal sarcoma in European dogs. *J Nat Cancer Inst, 37*:787, 1966.

Benirschke, K.; Malouf, N.; Low, R. J.; Heck, H.: Chromosome complement: differences between Equus caballus and Equus przewalskii, Poliakoff. *Science, 148*:382, 1965.

Chiarelli, B.: Chromosome polymorphism in the species of the genus Cercopithecus. *Cytologia, 33*:1, 1968.

Gropp, A.: Cytologic mechanisms of karyotype evolution in insectivores. In Benirschke, K. (Ed.): *Comparative Mammalian Cytogenetics*. Springer-Verlag, New York 1969, p. 247

Gustavsson, I., and Rockborn, G.: Chromosome abnormality in three cases of lymphatic leukemia in cattle. *Nature, 203*:990, 1964.

Hughes, D. T.: Cytogenetical polymorphism and evolution in mammalian somatic cell populations *in vivo* and *in vitro*. *Nature, 217*:518, 1968.

Koulischer, L.: Concept of cellular clonal evolution of karyotypes applied to evolution of species. *Acta Zool Path Antwerp, 48*:21, 1968.

Koulischer, L.; Tijskens, J.; Mortelmans, J.: Mammalian cytogenetics. I. The chromosomes of three species of Bovoidea: Bos taurus, Bison bonasus and Cephalophus grimmi. *Acta Zool Path Antwerp, 43*:135, 1967.

Lejeune, J.; Leucémies et cancer. In Turpin, R., and Lejeune, J.: *Les Chromosomes Humains*. Gauthier-Villars éd., Paris 1965.

Levan, A.: Some current problems of cancer cytogenetics. *Hereditas, 57*:343, 1967.
Levan, G.: Contribution to the chromosomal characterization of the PTKI rat kangaroo cell line. *Hereditas, 64*:85, 1970.
Makino, S.: Some epidemiologic aspects of venereal tumors of dogs as revealed by chromosome and DNA studies. *Ann NY Acad Sci, 108*:1106, 1963.
Matthey, R.: Les chromosomes des vertébrés. Librairie de l'Université F. Rouge, Lausanne, 1949 (1 vol.).
Muldal, S.; Elejalder, R.; Harvey, P. W.: Specific chromosome anomaly associated with autonomous and cancerous development in man. *Nature, 299*:48, 1971.
Ohno, S.: *Evolution by Gene Duplication.* Springer-Verlag, Heidelberg-New York, 1970, Vol. 1.
Robertson, W.: Chromosome studies. I. Taxonomic relationships shown in the chromosomes of Tettigidae and Acrididae. V-shaped chromosomes and their significance in Acrididae, Locustidae and Gryllidae: Chromosomes and variation. *J Morph, 27*:179, 1916.
Taylor, K. M.; Hungerford, D. A.; Snyder, R. L.; Ulmer, F. A.: Uniformity of karyotypes in the Camelidae. *Cytogenetics, 7*:8, 1968.
Weber, W. T.; Nowell, P. C.; Hare, W. C. D.: Chromosome studies of a transplanted and a primary canine venereal sarcoma. *J Nat Cancer Inst, 35*:537, 1965.
Wurster, D. H.: Cytogenetic and phylogenetic studies in Carnivora. In Benirschke, K.: *Comparative Mammalian Cytogenetics,* Heidelberg- New York, Springer-Verlag, 1969.
Wurster, D. H., and Benirschke, K.: Chromosome studies in the superfamily Bovoidea. *Chromosoma, 25*:152, 1958.
Wurster, D. H.; Benirschke, K.; Noelke, H.: Unusually large sex chromosomes in the sitatunga (Tragelaphus spekei) and the blackbuck (Antilope cervicapra). *Chromosoma, 23*:317, 1968.

Chapter VI

THE USE OF TECHNIQUES REVEALING CONSTITUTIVE HETEROCHROMATIN

IN 1970 and 1971, new techniques engendered excitement in the field of human cytogenetics. These were fluorochrome staining and the Giemsa-constitutive heterochromatin staining methods. These methods have brought better understanding of distribution of different kinds of DNA along the length of the chromatid allowing better identification and understanding of chromosome constitution. It now appears that chromosomes of eukaryotic species, including man, are chiefly composed of at least two types of DNA:

1. The bulk, or main DNA, carrying most of the genetic information and containing unique nucleotide sequences. It is sometimes designated as main band DNA (on the basis of buoyant density gradient centrifugation).

2. A rapidly reassociating DNA called satellite or repetitive DNA (composing constitutive heterochromatin).

Satellite DNA differs in base composition from the rest of nuclear DNA, consisting of families of highly multiplicated, repeated, nucleotide sequences. It possibly carries out only a limited number of functions. Despite the finding that the part of the chromosome which contains satellite DNA also contains small amounts of other types of DNA (Eckhardt and Gall, 1971), this particular area may be visualized by special techniques demonstrating *constitutive heterochromatin*. In contrast to the heteropyknotic X-chromosome in females designated as facultative heterochromatin, constitutive heterochromatin is not a variable state of chromatin. It is found in specific chromosomes throughout ontogenic development and is present in homologous autosomes in identical location. It varies, however, in its degree of condensation in cells of different types (Lee and Yunis, 1971). Some of the constitutive heterochromatin expresses itself as highly condensed chromocenters in interphase nuclei, and also most likely assumes a certain role in formation of nucleoli. Considerable amount of constitutive heterochromatin in human chromosomes occupies centromeric areas and forms bands made visible by fluorescent methods (Caspersson et al., 1970) and techniques based on treatment *in vitro* with sodium hydroxide, followed by incubation in sodium chloride, trisodium citrate and other modifications including trypsin treatment (Arrighi and Hsu, 1971; Sumner et al., 1971; Yunis et al., 1971; Drets and Shaw, 1971; Wang and Fedoroff, 1972; Seabright, 1972; and others). Wide implications of these methods are suggested mainly in precise identification of human chromosomes, better understanding of the cell cycle, and chromosome function (See Figs. 5, 6, 7).

It seems obvious that research on satellite DNA distribution and its role in chromosomes of malignant tumors and leukemias may shed new light on the mechanism of malignant transformation.

One of the avenues of investigation has been the search for a "common specific chromosomal pathway" for the origin of human malig-

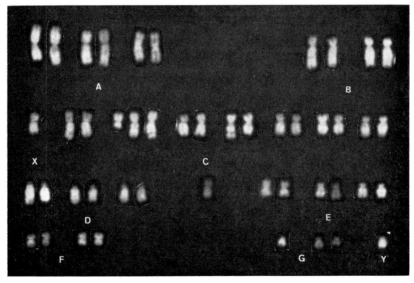

Figure 4. Fluorescent study of the karyotype of the P3 JHRIK Burkitt cell line. Anomalies are the following: monosomy 21, a supernumerary chromosome the size of a small C, and partial trisomy for the short arms of a C7. In this cell line, two C chromosomes with secondary constrictions were observed: one belongs to pair 7 and the other to pair 9. (Courtesy of Dr. Petit, Department of Pathology, University of Brussels.)

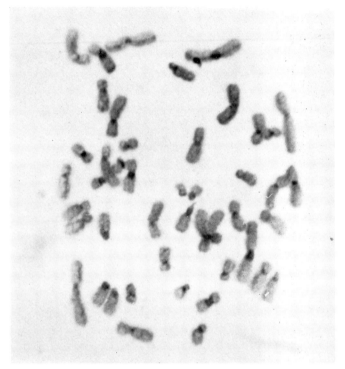

Figure 5. Human metaphase chromosomes treated with NaOH at pH 7 for two minutes followed by incubation in 6 × SSC at 65°C overnight to demonstrate centromeric constitutive heterochromatin. The largest blocks are in chromosomes A_1, C_9 and E_{18}.

nancy already proposed by Boveri in 1902, and rather unconvincingly pursued by several authors thereafter. Gofman et al. (1967) and Minkler et al. (1971) investigated the frequency of chromosome E_{16} in human malignancies and in cell lines. Their statistical data suggested that the number of chromosomes E_{16} per malignant karyotype is significantly elevated. These authors also assume the possibility that not only absolute elevation of E_{16}, but also the $E_{16}/G+Y$ ratio imbalance represents a common pathway for the development of human cancer in a certain proportion of cases, but excluding, for example, chronic myelogenous leukemia and Burkitt's lymphoma. The new method of identification of centromeric constitutive heterochromatin shows that chromosome E_{16} carries distinctly larger blocks of near-

Techniques Revealing Constitutive Heterochromatin

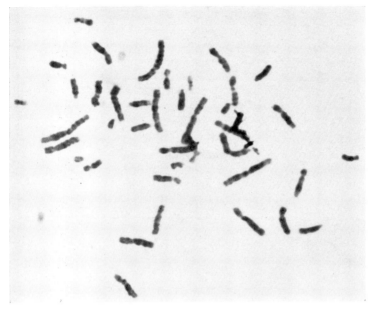

Figure 6. Human metaphase chromosomes treated with NaOH at pH 7 for two minutes followed by incubation in Sörensen buffer overnight to demonstrate G-bands, i.e. satellite DNA in the form of transverse bands of different density and extent.

centromere located satellite DNA than any other chromosome except A_1 and C_9. It is apparent then, that this technique would enable identification of E_{16} in some cases when its actual presence might not be apparent because of its changed morphology by deletion or translocation, or because it constitutes part of a large marker chromosome.

By the same method, and more easily by methods used for demonstrating constitutive heterochromatin in the form of banding patterns, the identity of a number of markers could be determined in the future, as well as the true origin of various supernumerary chromosomes. More insight could be gained into the problem of minute chromosomes (microchromosomes), their centromeres, and their replication pattern.

By using the above methods, attempts could be made to better understand the mechanism of the effect of radiation therapy and chemotherapy discovering possible sites of predilection on the chromosomes. Certainly numerous other implications of these techniques are to be

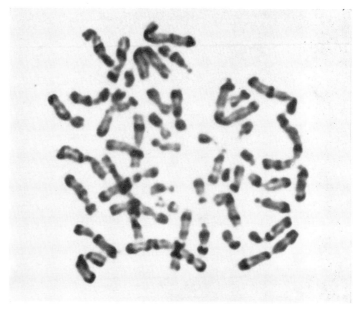

Figure 7. Human metaphase chromosomes treated by 0.025% trypsin solution in 0.02% EDTA for two minutes at 37°C followed by incubation in Sörensen buffer for five minutes to demonstrate telomeric constitutive heterochromatin.

expected, especially since the recent work of Smith, Maio, and others with virus SV_{40} transformation. It has been demonstrated that mouse satellite DNA is among the first cellular DNA which is induced to replicate after infection with polyoma virus and not at later times, as is usual in tissue culture where mouse cells are induced to replicate their DNA by exposure to culture media and serum. Furthermore, Maio (1971) reported two stable cell lines from the African green monkey: one lacking a satellite fraction, the other containing satellite DNA. Working with these two cell lines, Ritzi and Levine (1970) reported that cellular DNA synthesis is not induced by SV_{40} infection in the cell line lacking satellite DNA, but was rapidly induced in the other one containing satellite fractions.

It now seems that this avenue of research may provide additional information, not only about the function of satellite DNA in cell replication, but also about its possible role in malignant transformation.

Several other teams reported experiments indicating that malignancy can be suppressed by changing the chromosomal constitution of a cell. Pollack et al. (1970), for example, showed that when 3T3 cells transformed by SV_{40} virus revert to a nonmalignant phenotype their median chromosome number increases from 75 to 118. Rabinowitz and Sachs (1970) similarly observed the reversion to normal of hamster embryo cells transformed by polyoma virus being correlated with an increased chromosome number. From this, the authors postulated that the balance between two kinds of chromosomes—those carrying factors for expression, and those with factors for suppression of transforming genes—determines whether or not the malignant phenotype is expressed (Hitotsumachi et al., 1971). It now appears that reversion can be associated with an increase as well as a decrease in chromosome number. It will not be surprising to find imbalance of two particular groups of chromosomes being responsible for either reversion to normal or malignant transformation. If this hypothesis is to be validated, the importance of the new methods for identification of chromosomes discussed above is apparent.

REFERENCES

Arrighi, F. E., and Hsu, T. C.: Localization of heterochromatin in human chromosomes. *Cytogenetics, 10*:81, 1971.

Caspersson, T.; Zech, L.; Johansson, C. et al.: Identification of human chromosomes by DNA-binding fluorescent agents. *Chromosoma, 30*:215, 1970.

Drets, M. E., and Shaw, M. W.: Specific banding patterns of human chromosomes. *Proc Nat Acad Sci, 68*:2073, 1971.

Eckhart, R. A., and Gall, J. G.: Satellite DNA associated with heterochromatin in Rhynchosciara. *Chromosoma, 32*:407, 1971.

Gofman, J. W.; Minkler, J. L.; Tandy, R. K.: A specific common chromosomal pathway for the origin of human malignancy. *U of Calif Lawrence Rad Lab* Ref 50356, 1967.

Hitotsumachi, S.; Rabinowitz, Z.; Sachs, L.: Chromosomal control of reversion in transformed cells. *Nature, 231*:511, 1971.

Lee, J. C., and Yunis, J. J.: A developmental study of constitutive heterochromatin in Microtus agrestis. *Chromosoma, 32*:237, 1971.

Maio, J. J.: DNA strand reassociation and polyribonucleotide binding in the African green monkey, Cercopithecus aethiops. *J Mol Biol, 56*:579, 1971.

Minkler, J. L.; Gofman, J. W.; Tandy, R. K.: A specific common chromosomal pathway for the origin of human malignancy—II. *Br J Cancer, 24*:726, 1971.

Pollack, R.; Wolman, S.; Vogel, A.: Reversion of virus-transformed cell lines: hyperploidy accompanies retention of viral genes. *Nature, 228*:938, 1970.

Rabinowitz, Z., and Sachs, L.: Control of reversion of properties in transformed cells. *Nature, 225*:136, 1970.

Ritzi, E., and Levine, A. J.: Deoxyribonucleic acid replication in simian virus 40-infected cells. *J Virol, 5*:686, 1970.

Seabright, M.: The use of proteolytic enzymes for the mapping of structural rearrangements in the chromosomes of man. *Chromosoma 36*:204, 1972.

Sumner, A. L.; Evans, H. J.; Buckland, R. A.: New technique for distinguishing between human chromosomes. *Nature New Biol, 232*:31, 1971.

Walker, P. M. B., and McLaren, A.: Fraction of mouse deoxyribonucleic acid on hydroxyapatite. *Nature, 208*:1175, 1965.

Wang, H. C. and Fedoroff, S.: Banding in human chromosomes with trypsin. *Nature New Biol 235*:52, 1972.

Yunis, J. J.; Roldan, L.; Yasmineh, W. G.; Lee, J. C.: Staining of satellite DNA in metaphase chromosomes. *Nature 231*:532, 1971.

PART II
HEMATOPOIETIC AND LYMPHATIC SYSTEMS

Chapter VII

CHRONIC MYELOID (GRANULOCYTIC) LEUKEMIA

CYTOGENETICALLY, leukemias are the best documented forms of neoplasia in man. The reasons are numerous, the main one being the relative accessibility of the neoplastic tissue. The second is the relatively simple use of conventional cytogenetic techniques. For example, complicated dissociation of cells as in the case of solid tumors is not required.

Among blood disorders, chronic myeloid leukemia (CML) occupies a unique position, being the first example of neoplasia showing a specific chromosome marker (Nowell and Hungerford, 1960). Because of the large number of chromosome studies that have been concerned with this problem, a special chapter is devoted to cover the present knowledge concerning the cytogenetics of CML.

THE PHILADELPHIA CHROMOSOME—Ph^1

In two cases of CML, Nowell and Hungerford (1960) described a karyotype with 46 chromosomes including one G group chromosome which was smaller than the remaining three. The presence of the small marker was soon confirmed by Baikie et al. (1960) and again by Nowell and Hungerford (1961). Morphologic appearance of the chromosome is such that it might have arisen either from deletion or translocation, a question still not resolved (Caspersson et al., 1971). Following the Denver system of nomenclature (1960), which recommended that each new specific human chromosome marker should be identified by the two first letters of the city in which it had been discovered, the small G group marker was called the Philadelphia or Ph chromosome. Since it was the first marker chromosome to be discovered in Philadelphia, the suffix 1 followed the two letters, "Ph^1."

Since 1961, many workers have observed the Ph^1 chromosome in

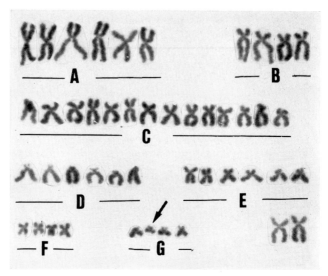

Figure 8. Chronic myeloid leukemia in a female patient with a characteristic Ph^1— "Philadelphia" chromosome (arrow). (Courtesy of Dr. Verhest, Department of Pathology, Institut Jules Bordet, Brussels.)

CML. As stated above, CML became the first and still unique example of a neoplasia with a specific chromosome marker. The Ph^1 chromosome brought substantial support to the concept of chromosome aberration as the primary event in carcinogenesis, thus greatly enhancing cytogenetic research in the field of cancer.

In a review of 636 reported cases (De Nava, 1970), it was shown that more than 90 per cent of all patients with CML exhibit the Ph^1 (Table V.). The specificity of the marker thus appears to be proven.

Localization of the Ph^1 Chromosome
Importance of Bone Marrow Studies

The Ph^1 chromosome is found exclusively in cells of the hematopoietic system. Other tissues, such as fibroblasts from skin or aponeurosis, do not show the Ph^1 chromosome (Baikie et al., 1959; Fitzgerald, et al., 1963). There is evidence that Ph^1 positive cells are derived from granulocytic, erythrocytic, and megakaryocytic stemlines. Thus, Whang et al. (1963) observed in 11 patients that *all* mitoses of bone marrow cells exhibited a Ph^1 chromosome. Bone marrow smears of the same

TABLE V
KARYOTYPE FINDINGS IN 636 REPORTED CASES
OF CHRONIC MYELOID LEUKEMIA

Karyotype	45,X,Ph¹,Y−	46, Ph¹	46, Ph¹ (pseudo-diploid)	46, no Ph¹	Others with Ph¹	Others without Ph¹
Number of Cases	12	486	15	50*	68	5
%	0.8	77.0	2.2	7.8	10.6	0.6

*Bone marrow studied in only 21 cases.

patients showed 25 to 60 percent erythroid cells, some in mitosis. The authors concluded that erythroblasts also had the Ph^1 chromosome; similar observations were made by Trujillo and Ohno (1963), Tough et al. (1963) and Fitzgerald et al. (1963). Clein and Flemans (1966) gave direct support to these statistical assumptions. They demonstrated histochemically the presence of iron granules in mitoses with the Ph^1, thus proving that these were dividing erythroblasts.

Highly polyploid mitoses may represent megakaryocytes (Sandberg et al., 1962; Tough et al., 1963; Whang et al., 1963). One Ph^1 chromosome in these polyploid cells was observed for each diploid set of chromosomes. Thus all three stemlines of the hematopoietic system involved in CML may be Ph^1 positive.

It is clear that the only reliable observations in CML are those derived from bone marrow cells *in situ* or circulating in the blood. In the cases in which peripheral blood only was studied, it was shown that when the white cell count was lower than 20,000/mm^3, there was not much chance of detecting the Ph^1 chromosome. Furthermore, when circulating lymphocytes were stimulated by phytohemagglutinin, as in routine blood cultures, they also proved to be without the Ph^1 chromosome. The best technique is incubation of blood for 24 to 48 hours *in vitro* without phytohemagglutinin. While mature lymphocytes will not divide, circulating marrow cells such as myelocytes, promyelocytes, and myeloblasts (Speed and Lawler, 1964; Sandberg et al., 1964) will yield metaphase spreads.

Data presented in Table V reflect that in 29 Ph^1 negative cases, blood alone was studied. If only those cases were compiled in which bone marrow was studied, nearly 95 per cent of the cases of CML will be Ph^1 positive.

Identity of the Ph^1 Chromosome: G21 or G22?

The Ph^1 chromosome is an autosome which has lost about 40 percent of its long arms (Rudkin et al., 1964). In the past, there was general agreement that the Ph^1 was a G21 chromosome. Various arguments were used to support this view. For example, it is known that the risk of developing acute leukemia in a patient with trisomy 21 is increased 20-to 30-fold compared with individuals of the same age not exhibiting Down's syndrome (Krivit and Good, 1956, 1957;

Doll et al., 1962). Furthermore, lobulation of polymorphonuclear leukocytes in 21 trisomic patients is less than normal (Turpin and Bernyer, 1947). These findings have been interpreted as evidence for "leukopoietic genes" located on chromosome 21. An excess of genetic material, as in Down's syndrome, could alter the normal metabolism of leukocytes and favor, in some undetermined way, the onset of a leukemic process. Other arguments were supported by enzymology: leukocyte alkaline phosphatase (LAP) level of patients with trisomy 21 is elevated (Alter et al., 1962; Trubowitz et al., 1962) and reduced in patients with CML (Valentine, 1951). However, a simple gene-enzyme relationship is not likely. In trisomy 21, the LAP level is about 50 percent higher than normal; in CML it is around 0. On the other hand, the LAP level has been elevated in some patients with Ph^1 positive CML and also in a patient with ulcerative colitis (Teplitz et al., 1964).

Recently, use of quinacrine mustard for study of chromosome fluorescence has shown that the Ph^1 chromosome stains as a G22 and not as a G21 (See Figs. 9 and 10). It was convincingly demonstrated by comparison of CML and trisomic 21 patients (Caspersson et al., 1971; O'Riordan et al., 1971). The history of attempts to identify the Ph^1 chromosome demonstrates the difficulties of localizing genes on chromosomes by means of combined enzymatic and cytogenetic techniques in man.

Specificity of the Ph^1 Chromosome

Table V shows that more than 90 percent of all cases of CML are Ph^1 positive. However, there are Ph^1 negative cases and the specificity of this marker in CML needs further elucidation.

Twin Studies and the Ph^1 Chromosome

Three pairs of monozygotic twins have been studied in which one of the twins was affected by CML while the other was healthy (Goh and Swisher, 1965; Dougan et al., 1966; Jacobs et al., 1966). The Ph^1 chromosome was observed only in bone marrow cells of the twin with CML. These reports demonstrate that the Ph^1 anomaly is acquired, not inherited, and that it is associated with the disease.

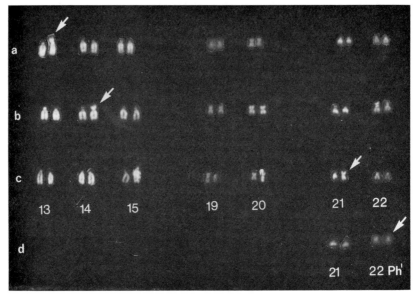

Figure 9. Partial karyotypes from four patients: (a) 46, XX, 13−, t(13q 21q) +; (b) 46, XX, 14−, t(14q 21q) +; (c) 46, XX, 21−, t(21q 21q) +; or 46, XX, 21qi; (d) G group chromosomes from a patient with CML and Ph[1] chromosome demonstrated as number 22 by less pronounced fluorescence compared with number 21. (Courtesy of Dr. M. L. O'Riordan et al.: Nature, 230:167, 1971.)

CML Without the Ph[1] Chromosome

To avoid complications, only those cases in which bone marrow cells were studied shall be considered here. Cases of CML without the Ph[1] chromosome have been reported rarely: three patients of Block et al. (1963), one patient of Speed and Lawler (1964), two of 27 patients of Tough et al. (1963). Exceptional are the following: 13 of 73 cases of Tjio et al. (1966) and 28 of 107 patients of Sandberg et al. (1971). This latter series apparently includes the Ph[1] negative cases published by Sandberg et al. in 1962 and Krauss et al. in 1964.

In childhood, CML is rare; Iverson (1966) found only seven cases among 516 leukemic children, Reisman and Trujillo (1963) found seven in 160 leukemic patients under ten years of age. The Ph[1] chromosome was not always found (Reisman and Trujillo, 1963;

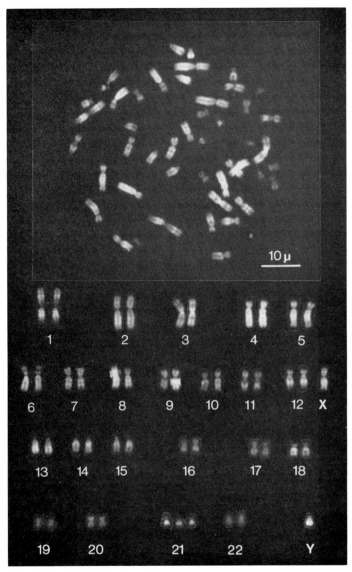

Figure 10. Metaphase spread and karyotype of trisomy 21 (Down's syndrome) patient. Note that the trisomic G_1 chromosome belongs to the brighter fluorescing pair when compared with Ph^1 chromosome in Figure 9. (Courtesy of Dr. M. L. O'Riordan et al.: Nature, 230:167, 1971.)

Hardisti et al., 1964; Tijo et al., 1966; Holton and Johnston, 1968).

Tjio et al. (1966) concluded that the clinical course of Ph^1 negative CML is different from that of the Ph^1 positive disease. Survival time is shorter (18 months vs. 45 months), blood cell count is not typical, and response to usual treatment is poorer. It is therefore tempting to propose that Ph^1 negative cases may form a separate clinical entity: the term chronic myeloid leukemia should then be utilized only for the cases with the Ph^1 chromosome (Kenis and Koulischer, 1967).

Ph^1 Chromosome Without CML

A Ph^1 or G chromosome identical to the Ph^1 chromosome has been observed in conditions other than CML. However, these observations are not frequent.

A few reports concern myeloproliferative disorders (MD). A Ph^1 chromosome has been noticed in two patients with thrombocythemia and platelet counts ranging from 900,000 to 4,000,000/mm^3 (Tough et al., 1963; Heath and Moloney, 1965), and in cases of polycythemia vera (PV) (Kemp et al., 1961, 1964; Levin et al., 1967; Koulischer et al., 1967).

One case of Kemp et al. is of special interest. The patient exhibited typical polycythemia vera until 1961. When cytogenetic analysis was performed, the total hemoglobin had dropped and a Ph^1 chromosome was observed. Some months later, the patient developed Ph^1 positive characteristic CML. This report indicates that the Ph^1 chromosome may be present before any clinical manifestations of CML. Further discussion concerning this problem will be found in Chapter I. (Chromosomes and premalignancy).

A Ph^1 chromosome has also been observed in cases of acute myeloblastic leukemia (AML) or chronic erythroleukemia. Hosfeld et al. (1971) reported three new cases—two AML and one chronic erythroleukemia—and reviewed nine other similar reports (Borges et al, 1962; Baguena Candela et al., 1964; Castoldi et al., 1964; Fortune et al., 1962; Grossbard et al., 1968; Kahn and Martin, 1967; Kiossouglou et al., 1968; Luers et al., 1962; Mastrangelo et al., 1967). It is noteworthy that nearly all reports showing a Ph^1 chromosome without clinical CML concern hematologic disorders somewhat related to CML. These cases seem to justify the conclusion of Hosfeld et al. (1971) that "the finding of a Ph¹ chromosome does not make a diagnosis." In our opinion it would be better to add ". . . in 2-3% of all Ph¹ positive cases."

CYTOGENETICS OF THE BLASTIC CRISIS

From a cytogenetic viewpoint, great uniformity characterizes the chronic phase of myeloid leukemia. The vast majority of reported cases manifests a pseudodiploid karyotype with 46 chromosomes including the Ph¹ chromosome. This uniformity disappears during the blastic crisis. Aneuploidy is often observed with a marked tendency towards hyperdiploidy (Lawler and Galton, 1966). With few exceptions (Tough *et al.*, 1961; Hungerford and Nowell, 1962), the Ph¹ chromosome is still present. However, aneuploidy is not necessary for the onset of a blastic crisis. Many cases of blastic crisis are known in which there is a 46, Ph¹ karyotype. Chronic and blastic phases of the disease thus cannot be distinguished using cytogenetic methods alone. Nevertheless, blastic crisis of acute myeloid leukemia is undoubtedly associated with karyotype instability which will be discussed further. Clinical implications will be considered in another section.

Blastic Crisis of CML and the Question of Specific Karyotype Alterations

A series of 84 cases of blastic crisis in which one or more aneuploid cell lines were observed will be analyzed in the present discussion. For the sake of clarity, references will be given later. It is obvious (Table VI) that more than 90 percent of all aneuploid cell lines have hyperdiploid counts while only 9.4 percent exhibit hypodiploidy. Most studies show that chromosomes of all groups are involved. However, (Fig. 11) gains of chromosomes in groups C and G, as well as duplication or triplication of the Ph¹ chromosome are more commonly observed than supernumerary chromosomes in groups D, E, and F. The large chromosomes of groups A and B, as well as marker chromosomes other than the Ph¹ chromosome, appear to be exceptional.

Duplication of the Ph¹ Chromosome

Duplication of the Ph¹ chromosome is the most frequent observation (Table VII). When more than 47 chromosomes are counted,

TABLE VI
CHROMOSOME COUNTS IN 84 CASES OF CML SHOWING ANEUPLOID CELL LINES AND STUDIED DURING THE BLASTIC CRISIS
(Review of the literature)

No. of Chrom.	44	45	47	48	49	50	51	52	53	55	58
No. of Cases	1	7	39	10	13	4	1	4	3	1	1
% of Total	1.1	8.3	46.4	11.9	15.4	4.8	1.1	4.8	4.1	1.1	1.1

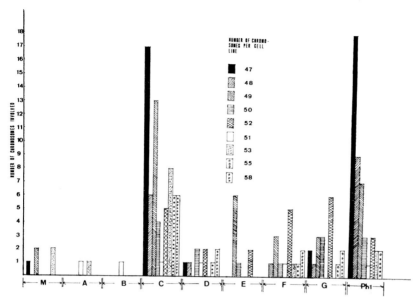

Figure 11. Distribution of chromosomes by Denver group in CML in blastic crisis of CML. Each chromosome number characteristic of a cell line is represented by a different shadowing of the columns. The height of each column is proportional to the number of chromosomes involved in the hyperdiploidy. Each group of chromosomes has been considered separately. Thus, in 39 cases with a count of 47 (black column), a supernumerary C has been observed 17 times, a supernumerary D only once, different markers once, a supernumerary G twice, and duplication of the Ph^1 18 times.

very often a supernumerary C group chromosome is observed. It is not known yet which one of the eight pairs of the C group is involved in each case. New available techniques such as fluorescent staining or study of banding patterns will help to solve this important question. Should always the same C group chromosome be involved, a new specific pattern of chromosome evolution in the blastic crisis of CML might be isolated. The Ph^1 chromosome often shows a tendency to duplicate or even to triplicate. Although the mechanism is not known, either nondisjunction or selective endoreduplication may be the cause (see Fig. 12).

Other Hyperdiploid Karyotypes

Hyperdiploidy without duplication of the Ph^1 chromosome has been

TABLE VII
KARYOTYPES OBSERVED DURING THE BLASTIC CRISIS OF CML

No. of Chrom.	Karyotype	Authors
46	Ph¹, Ph¹, C+, F+, G+, A—, B—, D—, E—	Pergoraro et al. (1967) Nowell and Hungerford (1961)
47	Ph¹ Ph¹	Hammouda et al. (1964) Hampel (1964), Kiossouglou et al. (1964); Dougan and Woodliff (1965); Ruffie et al. (1965); Rigo et al. (1966); Smalley (1966); Tjio et al. (1966); Duvall et al. (1967); Kenis and Koulischer (1967); De Nava (1970).
48	Ph¹, Ph¹, C+	Castro-Sierra et al. (1967); Lawler and Galton (1966); De Nava (1970)
	Ph¹, Ph¹, C+, C+, E—	Tjio et al. (1966)
	Ph¹, Ph¹, C+, 17—, mar+	De Nava (1970)
49	Ph¹, Ph¹, C+	Lawler and Galton (1966)
	Ph¹, Ph¹, C+, C+, 17qi	Stich et al. (1966)
	Ph¹, Ph¹, F+, G+	Erkmann et al. (1966)
50	Ph¹, Ph¹, C+, C+, C+	De Nava (1970)
	Ph¹, Ph¹, C+, C+, C+, 17—, mar+	id.
	Ph¹, Ph¹, C+, 16+, 16+	id.
	Ph¹, Ph¹, C+, D+, F+	Engel and McKee (1966)
	Ph¹, Ph¹, D+, E+, G+	id. Hammouda et al. (1964)
51	Ph¹, Ph¹, C+, B+, D+, E+	Ilbery and Louer (1966)
	Ph¹, Ph¹, C+, A+, B+, D+, F+	Kemp et al. (1964)
52	Ph¹, Ph¹, C+, C+, F+, G+, G+	Castro-Sierra et al. (1967)
	Ph¹, Ph¹, C+, D+, D+, F+, G+, 17—, mar+	De Nava (1970)
	Ph¹, Ph¹, 18+, 18+, 19+, 19+, G+	id.
53	Ph¹, Ph¹, C+, C+, C+, F+	Pergoraro et al. (1967)
	Ph¹, Ph¹, 6C+, A+, G+, E—, mar+	Pedersen and Videbaek (1964)
55	Ph¹, Ph¹, 6C+, D+, F+, G+	Erkmann et al. (1966)
58	Ph¹, Ph¹, 6C+, 2D+, 2F+, 2G+	Erkmann et al. (1966)
48	Ph¹, Ph¹, Ph¹	deGrouchy et al. (1965); Schroeder and Böök (1965)
	Ph¹, Ph¹, Ph¹, D+, G—	Streiff et al. (1966)

Note: All karyotypes listed in this table exhibit common duplication of the Ph¹ chromosome.

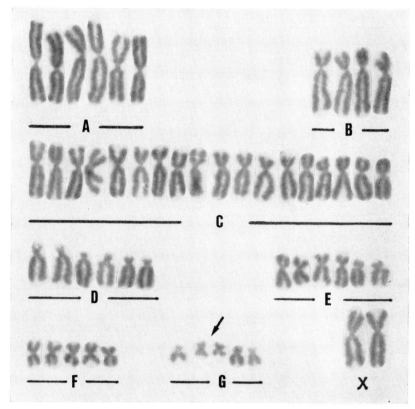

Figure 12. Blastic crisis of chronic myeloid leukemia in a female patient. There are, in addition to duplication of the Ph^1 chromosome (arrow), 4 supernumerary chromosomes in group C and one in group F (52, XX, Ph^1 Ph^1+, 4 C+, F+). (Courtesy of Dr. Verhest, Department of Pathology, Institut Jules Bordet, Brussels.)

repeatedly observed during the blastic crisis of CML (Table VIII). Supernumerary C group chromosomes are frequent, the simplest karyotype in this group being 47, Ph^1, C+. Apparently, similar karyotypes have been observed by different authors. For instance, the karyotype 47, Ph^1, C+, C+, E— has been shown in patients of Court Brown and Tough (1963), Pedersen (1964), and Krauss et al. (1966); the karyotype 49, Ph^1 C+E+G+ has been reported by Court Brown and Tough (1963), Lejeune et al. (1965), and Kenis and Koulischer (1967). While possibly mere coincidence, it could also

TABLE VIII

HYPERDIPLOID KARYOTYPES OBSERVED DURING THE BLASTIC CRISIS OF CML

No. of Chrom.	Karyotype	Authors
47	Ph¹, C+	Court Brown and Tough (1961); Goh et al. (1964); Kiossouglou et al. (1965); Engel and McKee (1966); De Nava (1970).
47	Ph¹, C+, C+, E—	Court Brown and Tough (1961); Pedersen (1964); Krauss et al., (1966), De Nava (1970).
47	Ph¹, G+	Stahl et al. (1966)
48	Ph¹, C+, F+	Wahrman et al. (1967)
48	Ph¹, C+, D+, mar+, 17—	De Nava (1969)
49	Ph¹, C+, E+, G+	Court Brown and Tough (1963; Lejeune et al. (1965); Kenis and Koulischer (1967)
	Ph¹, 18+, 19+, G+	De Nava (1969)
53	Ph¹, 5C+, A+, G+, mar+, E—	Pedersen (1965)

mean that some common patterns of chromosome evolution exist during the blastic crisis. Quite a consistent mode of clonal evolution has been proposed by de Grouchy et al. (see De Nava, 1969). Thirteen patients possessed a Ph¹ chromosome, loss of a 17-18 chromosome, and had a supernumerary C which might have been an isochromosome for the long arms of a 17 chromosome. Such a 17qi has also been observed by Engel et al. (1967). One cannot discount the possibility that chromosome evolution during blastic crisis may follow specific patterns and is not just due to chance or to variable environmental factors.

Hypodiploidy in Blastic Crisis

Hypodiploidy is not frequently observed during the blastic crisis (Table IX). Although possibly not directly relevant, it is interesting to note that loss of genetic material is also less common in congenital autosomal aberrations. Only a few deletion syndromes are known, autosomal monosomies being exceptional. Cells, as "whole organisms," may demonstrate the same inability to survive loss of part of the genetic material.

The case of Engel et al. (1965) merits some comment. Their 53-year-old female patient manifested two abnormalities in addition to

TABLE IX
HYPODIPLOID KARYOTYPES OBSERVED DURING THE BLASTIC CRISIS OF CML

No. of Chrom.	Karyotype	Authors
44	Ph^1, C—, E—	Kenis and Koulischer (1967)
	Ph^1, E—, F—	Court Brown and Tough (1963)
45	Ph^1, C—, C—, D+	Lawler and Galton (1966)
	Ph^1, t (G/D)	Ohno et al. (1961)
	Ph^1, D—C—, dic +	De Nava (1969)
	Ph^1, C—, F—, dic +	id
	Ph^1, 17—, 17—, mar +	id
	Ph^1, G—	Stahl et al. (1966)
	Ph^1, variable losses	Speed and Lawler (1964); Pedersen (1965) Tjio et al. (1966)

the Ph^1 chromosome, i.e. D/D translocation and an isochromosome for the long arms of a number 17 chromosome. The D/D translocation, however, was congenital and the hypodiploid number was thus not due to the leukemic process.

Chromosome Abnormalities With a Ph^1 Chromosome and a Diploid Number of 46 in Blastic Crisis

In some Ph^1 positive cases, the normal diploid number of 46 is maintained, but cytogenetic studies show obvious structural rearrangements (Table X). Frequently, karyotypes exhibit marker chromosomes, which have not been included in Figure 8. As already pointed out, new means of accurate identification of chromosomes involved in rearrangements should be employed before drawing conclusions about "common variants" (Lejeune, 1965).

TABLE X
PSEUDODIPLOID KARYOTYPES OBSERVED DURING THE BLASTIC CRISIS OF CML

No. of Chrom.	Karyotype	Authors
46	Ph^1, C+, 16—	Tough et al. (1961)
46	Ph^1, A—, mar +	Kiossouglou et al. (1965); Tjio et al. (1966)
46	Ph^1, C +, mar +, A—, 16—	Ford and Clarke (1963)
46	Ph^1, C—, mar +	Fitzgerald et al. (1966)
46	Ph^1, inv (Cp—q +)	Tough et al. (1963)
46	Ph^1, D+, 16+, C—, G—	De Nava (1970)
46	Ph^1, 17—, mar +	id

KARYOTYPES OTHER THAN 46, Ph¹ DURING THE CHRONIC PHASE OF THE DISEASE

Aneuploidy, that is, the presence of a cell line(s) with chromosome numbers other than 46, or pseudodiploidy with chromosome abnormalities other than the Ph^1 chromosome, are not necessarily exclusively linked with transformation of CML into blastic crisis. Several cases with gross structural rearrangements have been reported in patients during the chronic phase of the disease. Even duplication of the Ph^1 chromosome, so characteristic of the blastic crisis, has been reported in a patient in remission (Tjio et al., 1966). The karyotype (45,Ph^1,X,Y-) has been found in several male patients who have presented very slow evolution of the disease. This could be a particular form of CML, at least from the cytogenetic viewpoint.

CHROMOSOMES AND THERAPEUTICS

Do cytogenetic studies yield any information concerning the action of therapeutics on CML? Indeed, this is an important question.

The significance of a specific chromosome marker such as the Ph^1 chromosome or the presence of aneuploid cell lines is obvious. Any appropriate treatment should result in a decrease in the number of abnormal cells, and cytogenetics may be used to demonstrate clinical improvement. It has been shown that x-ray irradiation of the spleen has been associated with a decrease of Ph^1 positive cells in peripheral blood (Court Brown and Tough, 1963). When the leukocyte count is below 20,000/mm^3, Ph^1 positive cells are no longer observed in the blood, but are still present in the bone marrow. On the other hand, the disappearance of an aneuploid cell line in bone marrow may announce a remission or, at least, an improvement of the course of the disease (Goh et al., 1964). An interesting example of the application of genetic methods in problems of therapy has been given by Court Brown and Tough (1963). A 49, Ph^1 C+, E+, G+ cell line disappeared with 6-mercaptopurine (6MP) treatment. A new cell line appeared, with a 44, Ph^1, E—, F— karyotype. This line was resistent to 6MP and prednisone, but responded favorably to colcemid.

Theoretically, it should be possible to choose the therapeutic agent by following the development of the karyotype of the patient. Similar chromosome rearrangements imply similar genetic effects at the gene level, thus similar resistance or sensitivity to therapeutic agents. This is one reason why accurate study of karyotypes should be beneficial with the advent of better identification methods for chromosomes. Practical implications may result from this painstaking work, and perhaps shall be used as a model for other forms of human cancer. However, there are still simple related problems to be solved such as why a proportion of Ph^1 positive cells does not directly respond to treatment, and why the presence of Ph^1 cannot always be used as a prognostic aid concerning future evolution of the disease (Tjio et al., 1966; Lawler, 1967).

CONCLUSIONS

From the bulk of data accumulated since the initial discovery of the Ph^1 chromosome, a few general conclusions can be drawn. These perhaps may appear to be too schematic, but even at this stage they carry some importance. Certainly the reader should not get the impression that all the work done in this field until now is a catalog of odd karyotypes, of interest to only a few specialists:

1. Study of chromosomes in CML must be performed from bone marrow cells.

2. In 95 percent of all cases, the appearance of the Ph^1 chromosome suggests chronic myeloid leukemia. Thus, cytogenetics is very useful in establishing a definite diagnosis in borderline cases (myeloproliferative disorders).

3. Cytogenetics allows the identification of different clinical forms of the disease: "classic" CML with Ph^1 chromosome; "unusual" CML without Ph^1 chromosome (with poor response to usual therapeutic agents and fatal evolution in a short time); CML with loss of a Y-chromosome in males (very long disease course).

4. Aneuploidy or structural rearrangements are associated with acute exacerbation of the disease (blastic crisis in the great majority of cases). Some common pathways of chromosome evolution exist.

5. Currently, cytogenetics is not a therapeutic guide, but permits following the efficacy of a given treatment.

6. More work is needed using new methods, such as fluorescence

and banding techniques for revealing constitutive heterochromatin. Such methods based on denaturization of DNA followed by reassociation of satellite DNA to accurately identify the chromosomes involved in aneuploidy should aid immensely in this problem.

REFERENCES

Adams, A.; Fitzgerald, P. H.; Gunz, F. W.: A new chromosome abnormality in chronic granulocytic leukemia. *Br Med J, 2*:1474, 1961.

Alter, A. A.; Lee, S. L.; Pourfar, M.; Dobkin, G.; Leukocyte alkaline phosphatase in mongolism; a possible chromosome marker. *J Clin Invest, 41*:1341, 1962.

Atkin, N. B., and Taylor, M. C.: A case of chronic myeloid leukemia with a 45-chromosome cell-line in the blood. *Cytogenetics, 1*:97, 1962.

Baguena Candela, R., and Forteza Bover, G.: Estudio citogenetico de una leucemia aguda del tipo mieloblastico—promelocitico con cromosoma Filadelfia (Ph^1). *Med Esp, 50*:298, 1964.

Baikie, A. G.; Court Brown, W. M.; Jacobs, P. A.; Milne, J. S.: Chromosome studies in human leukemia. *Lancet, 2*:425, 1959.

Baikie, A. G.; Court Brown, W. M.; Buckton, K. E.; Harnden, D. G.; Jacobs, P. A.; Tough, I. M.: A possible specific chromosome abnormality in human chronic myeloid leukemia. *Nature, 188*:1165, 1960.

Block, J. B.; Carbone, P. P.; Oppenheim, J. J.; Frei, E.: The effect of treatment in patients with chronic myelogenous leukemia: biochemical studies. *Ann Intern Med, 59*:629, 1962.

Borges, W. H.; Wald, N.; Kim, J.: Non-specificity of chromosomal abnormalities in human leukemia. *Clin Res Proc, 10*:211, 1962 (abstr).

Caspersson, T.; Zech, L.; Johansson, C.; Modest, E. J.: Identification of human chromosomes by DNA-binding fluorescent agents. *Chromosoma* (Berl.), *30*:215, 1970.

Caspersson, T.; Gahrton, G.; Lindsten, J.; Zech, L.: Identification of the Philadelphia chromosome as a number 22 by quinacrine mustard fluorescence analysis. *Exp Cell Res, 63*:238, 1971.

Caspersson, T.; Hultén, M.; Lindsten, J.; Zech, L.: Distinction between extra G-like chromosomes by quinacrine mustard fluorescence analysis. *Exp Cell Res, 63*:240, 1971.

Castoldi, G.; Yam, L. T.; Mitus, W. J.; Crosby, W. H.: Chromosomal studies in erythroleukemia and chronic erythremic myelosis. *Blood, 31*:202, 1968.

Castro-Sierra, E.; Gorman, L. Z.; Merker, H.: Clinical and cytogenetic findings in the terminal phase of chronic myelogenous leukemia. *Humangenetik, 4*:62, 1967.

Clein, G. P.; Flemens, R. J.: Involvement of the erythyroid series in blastic crisis of chronic myeloid leukemia. Further evidence for the presence of Philadelphia chromosome in erythroblasts. *Br J Haemat, 12*:754, 1966.

Court Brown, W. M., and Tough, I. M.: Cytogenetic studies in chronic myeloid leukemia. *Advances Cancer Res, 7*:351, 1963.
De Nava, M. C.: Les anomalies chromosomiques au cours des hémopathies malignes et non malignes. *Monogr Ann Génét* (Paris), l'Expansion ed., 1969, Vol. 1.
Denver Conference: A proposed standard system of nomenclature of human mitotic chromosomes. *Acta Genet* (Basel), *10*:322, 1960.
Dougan, L., and Woodliff, H. J.: Presence of two Ph^1 chromosomes in cells from a patient with chronic granulocytic leukemia. *Nature, 205*:405, 1965.
Dougan, L.; Scott, I. D.; Woodliff, H. J.: A pair of twins, one of whom has chronic granulocytic leukemia. *J. Med Genet, 3*:217, 1966.
Duvall, C. P.; Carbone, P. P.; Bell, W. R.; Whang, J.; Tjio, J. H.; Perry, S.: Chronic myelocytic leukemia with two Philadelphia chromosomes and prominant peripheral lymphadenopathy. *Blood, 29*:652, 1967.
Engel, E.; McKee, B. J.; Hartmann, R. C.; Engel de Montmollin, M.: Two leukemic peripheral blood stemlines during acute transformation of chronic myelogenous leukemia in a D/D translocation carrier. *Cytogenetics, 4*:147, 1965.
Engel, E.; McKee, L. C.: Double Ph^1 chromosomes in leukaemia. *Lancet, 2*:337, 1966.
Engel, E.; McKee, L. C.; Bunting, K. W.: Chromosomes 17-18 in leukaemias. *Lancet, 2*:42, 1967.
Erkman, B.; Crookston, H. H.; Conen, P. E.: Double Ph^1 chromosomes in leukaemia. *Lancet, 1*:368, 1966.
Fitzgerald, P. H.; Adams, A.; Gunz, F. W.: Chronic granulocytic leukemia and the Philadelphia chromosome. *Blood, 21*:183, 1963.
Ford, C. E., and Clarke, M.: Cytogenetic evidence of clonal proliferation in primary reticular neoplasms. *Can Cancer Conf, 5*:129, 1963.
Fortune, D. W.; Lewis, F. J. W.; Poulding, R. H.: Chromosome pattern in myeloid leukaemia in a child. *Lancet, 1*:537, 1962.
Goh, K.; Swisher, S. N.; Troup, S. B.: Submetacentric chromosome in chronic myelocytic leukemia *Arch Intern Med, 114*: 439, 1964.
Goh, K.; Swisher, S. N.: Identical twins and chronic myelocytic leukemia. Chromosomal studies of a patient with chronic myelocytic leukemia and his normal identical twin. *Arch Intern Med, 115*:475, 1965.
Grossbard, L.; Rosen, D.; McGilvray, E.; DeCapoa, A.; Miller, O.; Bank, A.; Acute leukemia with Ph^1-like chromosome in an LSD user. *JAMA, 205*:791. 1968.
de Grouchy, J.; De Nava, C.; Bilski-Pasquier, G.: Duplication d'un Ph^1 et suggestion d'une évolution clonale dans une leucémie myéloide chronique en transformation aigue. *Nouv Rev Hémat Franc, 5*:69, 1965.
Hammouda, F.; Quaglino, D.; Hayhoe, F. G. S.: Blastic crisis in chronic granulocytic leukemia: cytochemical, cytogenetic and autoradiographic studies in four cases. *Br Med J, 1*:1272, 1964.

Hampel, K. E.: Diplo-Ph¹-Chromosom bei der myeloid Leukämie. 42:522, 1964.
Hardisty, R. M.; Speed, D. E.; Till, M.: Granulocytic leukaemia in childhood. Br J Haemat, 10:551, 1964.
Heath, C. W., and Moloney, W. C.: The Philadelphia chromosome in an unusual case of myeloproliferative disease. Blood, 26:471, 1965.
Holton, C. P., and Johnson, W. W.: Chronic myelocytic leukemia in infant siblings. J Pediat, 72:377, 1968.
Holland, W. W.; Doll, R.; Carter, C. O.: The mortality from leukaemia and other cancers among patients with Down's syndrome (mongols) and among their parents. Br J Cancer, 16:177, 1962.
Hossfeld, D. K.; Han, T.; Holdsworth, R. N.; Sandberg, A. A.: Chromosomes and causation of human cancer and leukemia. VII. The significance of the Ph¹ in conditions other than CML. Cancer, 27:186, 1971.
Houston, E. W.; Ritzman, S. W.; Levin, W. C.: Untreated chronic myelocytic leukemia associated with an unusual chromosome pattern. Ann Intern Med, 61:696, 1964.
Hungerford, D. A.; Nowell, P. C.: Chromosome studies in human leukemia. III. Acute granulocytic leukemia. J Nat Cancer Inst, 29:545, 1962.
Ilbery, P. L. T., and Lower, C. S.: Ph¹ chromosome in the differential diagnosis of a case of ascites. Aust Radiol, 10:135, 1966.
Iversen, T.: Leukemia in infancy and childhood. A material of 570 Danish cases. Acta Paediat Scand, 167(suppl):219, 1966.
Jacobs, E. M.; Luce, J. K.; Carllean, R.: Chromosome abnormalities in human cancer. Report of a patient with chronic myelocytic leukemia and his nonleukemic monozygotic twin. Cancer, 19:869, 1966.
Kemp, N. H.; Stafford, J. L.; Tanner, R. K.: *Cytogenetic Studies of Polycythemia Vera.* Vienna, 8th Congress Europ Soc Haemat, 1961, p. 192.
Kemp, N. H.; Stafford, J. L.; Tanner, R.: Chromosome studies during early and terminal chronic myeloid leukemia. Br J Med, 1:1010, 1964.
Kenis, Y., and Koulischer, L.: Etude clinique et cytogénétique de 21 patients atteints de leucémie myéloide chronique. Europ J Cancer, 3:83, 1967.
Khan, M. H., and Martin, H.: Myeloblastenleukämie mit Philadelphia-Chromosom. Klin Wschr, 45:821, 1967.
Kiossouglou, K. A.; Mitus, W. J.; Dameshek, W.: Two Ph¹ chromosomes in acute granulocytic leukemia. A study of two cases. Lancet, 2:665, 1965.
Koulischer, L.; Frühling, J.; Henry, J.: Observations cytogénétiques dans la maladie de Vaquez. Europ J Cancer, 3:193, 1967.
Krauss, S.; Sokal, J. E.; Sandberg, A. A.: Comparison of Philadelphia chromosome-positive and -negative patients with chronic myelocytic leukemia. Ann Intern Med, 61:625, 1964.
Krivit, W., and Good, R. A.: The simultaneous occurrence of leukemia and mongolism. Am J Dis Child, 91:218, 1956.
Krivit, W., and Good, R. A.: Simultaneous occurrence of mongolism and leukemia. Am J Dis Child, 94:289, 1957.

Lawler, S. D.: Chromosomes et transformation aigüe des leucémies myéloïdes chroniques. *Nouv Rev Franc Hemat*, 7:529, 1967.

Lawler, S. D., and Galton, D. A. G.: Chromosome changes in the terminal stages of chronic granulocytic leukemia. *Acta Med Scand*, 445(suppl):312, 1966.

Lejeune, J.; Berger, R.; Caille, B.; Turpin, R.: Évolution chromosomique d'une leucémie myéloïde chronique. *Ann Genet*, 8:44, 1965.

Levin, W. C.; Houston, E. W.; Ritzmann, S. E.: Polycythemia vera with Ph^1 chromosomes in two brothers. *Blood*, 30:503, 1967.

Luers, T.; Struck, E.; Boll, L.: Über eine spezifische Chromosomenanomalie bei Leukämie (das "minute" oder "Ph^1 Chromosom"). *München Wschr*, 10:1493, 1962.

Mastrangelo, R.; Zuelzer, W. W.; Thompson, R. I.: The significance of the Ph^1 chromosome in acute myeloblastic leukemia: serial cytogenetic studies in a critical case. *Pediatrics*, 40:834, 1967.

Nowell, P. C.; Hungerford, D. A.: A minute chromosome in human chronic granulocytic leukemia. *Science*, 132:1497, 1960.

Nowell, P. C., and Hungerford, D. A.: Chromosome studies in human leukemia. II. Chronic granulocytic leukemia. *J Nat Cancer Inst*, 21:1013, 1961.

O'Riordan, M. L.; Robinson, J. A.; Buckton, K. E.; Evans, H. J.: Distinguishing between the chromosomes involved in Down's syndrome (trisomy 21) and chronic myeloid leukemia (Ph^1) by fluorescence. *Nature*, 230:167, 1971.

Pedersen, B.: Two cases of chronic myeloid leukemia with presumably identical 47-chromosome cell-lines in the blood. *Acta Path Microbiol Scand*, 61:497, 1964.

Pedersen, B.: The aneuploid, Ph^1-positive cell population during progression and treatment of chronic myelogenous leukemia. *Acta Path Microbiol Scand*, 67:451, 1966.

Pedersen, B.; Videbaek, A.: Several cell-lines with abnormal karyotypes in a patient with chronic myelogenous leukemia. *Scand J Hemat*, 1:129, 1964.

Pegoraro, L.; Pileri, A.; Rovera, G.; Gavosto, F.: Trisomia del cromosoma Filadelfia (Ph^1) nella crisi blastica di un caso di leucemia mieloide cronica. *Tumori*, 53:315, 1967.

Reisman, L. E.; Trujillo, J. M.: Chronic granulocytic leukemia of childhood *J Pedatr*, 62:710, 1963.

Rigo, S. J.; Stannard, M.; Cowling, D. C.: Chronic myeloid leukemia associated with multiple chromosome abnormalities. *Med J Aust*, 2:70, 1966.

Rudkin, G. T.; Nowell, P. C.: DNA contents of chromosome Ph^1 and chromosome 21 in human chronic granulocytic leukemia. *Science*, 144:1229, 1964.

Ruffié, J.; Bierme, R.; Ducos, J.; Colombies, P.; Salles-Mourlan, A. M.; Sendrail, A.: Haplosome 21 associé a une trisome 6-12 chez un tuberculeun presentant une myeloblastose sanguine et medullaire. *Nouv Rev Franc Hemat*, 4:719, 1964.

Sanchez Cascos, A.; Barreiro, E.: Cromosoma Ph¹ y translocacion 5-12 en un caso de leucemia mieloide cronica. *Rev Clin Esp, 94*:10, 1964.

Sandberg, A. A.; Takaaki, T.; Crosswhite, L. H.; Hauschka, T. S.: Comparison of chromosome constitution in chronic myelocytic leukemia and other myeloproliferative disorders. *Blood, 20*:393, 1962.

Sandberg, A. A.; Kikuchi, Y.; Crosswhite, L. H.: Mitotic ability of leukemic leukocytes in chronic myelocytic leukemia. *Cancer Res, 24*: 1468, 1964.

Sandberg, A. A.; Hossfeld, D. K.; Ezdinli, E. Z.; Crosswhite, L. H.: Chromosomes and causation of human cancer and leukemia. VI. Blastic phase, cellular origin, and the Ph¹ in CML. *Cancer, 27*:176, 1971.

Schroeder, T. M.; Bock, H. E.: Trisomie des Ph¹-Chromosoms in Myeloblasten während der terminalen Phase einer chronisch myeloischen Leukämie. *Humangenetik, 1*:681, 1965.

Smalley, R. V.: Double Ph¹ chromosomes in leukemia. *Lancet, 2*:591, 1966.

Speed, D.; Lawler, S.: Chronic granulocytic leukemia: The chromosomes and the disease. *Lancet, 1*:403, 1964.

Stahl, A; Papy-Jurion, M. C.; Muratore, R.; Mongin, M.; Olmer, J.: Etude des chromosomes dans 35 observations de leucose. *Presse Med, 74*:429, 1966.

Stich, W.; Back, F.; Dormer, P.; Tsirimbas, A.: Doppel-Philadelphia-Chromosom und Isochromosom 17 in der terminalen Phase der chronischen myeloischen Leukämie. *Klin Wschr, 44*:334, 1966.

Streiff, R.; Peters, A.; Gilgenkrantz, S.: Anomalies chromosomiques au cours de la transformation blastique terminale d'une leucémie myéloide chronique. Prédominance d'un clone à 48 chromosomes avec 2 chromosomes Ph¹. *Nouv Rev Franc Hemat, 6*:417, 1966.

Teplitz, R. L.; Rosen, S. B.; Teplitz, M. R.: Granulocytic leukemia, Philadelphia chromosome and leukocyte alkaline phosphatase. *Lancet, 2*:418, 1964.

Tjio, J. H.; Carbone, P. P.; Whang, J.; Frei, E.: The Philadelphia chromosome and chronic myelogenous leukemia. *J. Nat Cancer Inst, 36*:587, 1966.

Tough, I. M.; Court Brown, W. M.; Baikie, A. G.; Buckton, K. E.; Harnden, D. C.; Jacobs, P. A.; King, M. J.; McBride, J A: Cytogenetic studies in chronic myeloid leukemia and acute leukemia associated with mongolism. *Lancet, 1*:411, 1961.

Tough, I. M.; Court Brown, W. M.; Baikie, A. B.; Buckton, K. E.; Harnden, D. G.; Jacobs, P. A.; Williams, J. A.: Chronic myeloid leukemia. Cytogenetic studies before and after splenic irradiation. *Lancet, 2*:115, 1962.

Tough, I. M.; Jacobs, P. A.; Court Brown, W. M.; Baikie, A. G.; Williamson, E. R. D.: Cytogenetic studies on bone-marrow in chronic myeloid leukemia. *Lancet, 1*:844, 1963.

Trubowitz, S.; Kirman, D.; Masek, B.: The leukocyte alkaline phosphatase in mongolism. *Lancet, 2*:486, 1962.

Trujillo, J. M.; Ohno, S.: Chromosomal alteration of erythropoietic cells in chronic myeloid leukemia. *Acta Haemat, 29*:311, 1963.

Turpin, R., and Bernyer, G.: De l'influence de l'hérédité sur la formule d'Arneth (cas particulier du mongolisme). *Revue d'Hemat,* 2:189, 1947.
Valentine, W. N., and Beck, W. S.: Biochemical studies on leukocytes. I. Phosphatase activity in health, leukocytosis and myelocytic leucemia. *J Lab Clin Med,* 38:39, 1951.
Wahrman, J.; Voss, R.; Shapiro, T.; Ashkenazi, A.: The Philadelphia 1 chromosome in two children with chronic myeloid leukemia. *Israel J Med Sci,* 3:380, 1967.
Whang, J.; Frei, E.; Tjio, J. H.; Carbone, P. P.; Brecher, G.: The distribution of the Philadelphia chromosome in patients with chronic myelogenous leukemia. *Blood,* 22:664, 1963.

Chapter VIII

ACUTE LEUKEMIAS

CYTOGENETIC STUDIES in different forms of acute leukemia (AL) have failed to demonstrate the presence of a specific marker such as the Ph^1 in CML. Moreover, a global view of the results concerning 482 different cases of AL (Table XI) shows that more than 53 percent of all patients studied had an apparently normal karyotype. Chromosome studied in AL may appear particularly disappointing; besides the absence of any specific marker, there is a wide range of chromosome aneuploidies in the same form of disease. This situation led Sandberg (1966) to conclude that, for him, "chromosomal abnormalities in leukemia and cancer in general described to date are secondary phenomena to the neoplastic state." Later, after statistical analysis, Sandberg et al. (1968) affirmed that "no particular group of chromosomes can be imputed as being responsible for the karyotypic changes." However, different opinions had been expressed, among others, by Van Steenis (1966), Levan (1966), Muldal et al. (1971), and, in fact, by Sandberg (1966) earlier.

TABLE XI
RESULTS OF CHROMOSOME STUDIES IN
482 DIFFERENT CASES OF AL
(Review of literature)

Ploidy	Hypo	Pseudo	Diplo	Hyper	Total
No. of Cases	53	31	258	143	482
%	11.0	6.3	53.2	29.5	100

Nevertheless, chromosome analysis in AL is of considerable interest if, instead of being placed in a general context, the results are considered at an individual level. Two views are then possible: clinical and fundamental.

From the clinical point of view, the presence of an abnormal karyotype offers a clear tool for study of the course of the disease. Thus,

Reisman et al. (1964) observed aneuploid cell lines in 13 children with AL during the acute phase of the disease. These lines disappeared after adequate treatment and reappeared, unchanged or slightly altered but still very recognizable, after relapse. Sandberg et al. (1968) made the same observation: "once a karyotypic picture is established in the leukemic cells, it does not appear to undergo any visible or major morphologic changes during various phases of the clinical course." This also is the opinion of Gunz et al. (1969) based on their study of 80 cases of AL. Certainly this persistence of abnormal karyotype supports the stemline concept.

The finding of aneuploidy has limited use as a prognostic aid in AL: Lampert (1968) studied 21 children with AL: 16 had acute lymphocytic leukemia (ALL), five acute myelocytic leukemia (AML), and one acute monocytic leukemia. Severe aneuploidy was found in six cases of ALL while all karyotypes and DNA nuclear values of AML were normal. Still, the mean survival period was 175 ± 51 days in AML and 817 ± 162 days in ALL.

From a fundamental point of view, study of individual cases is very important. For Levan (1969) "the chromosome variations in tumors is never haphazard, but gathers around stemlines and sidelines"; moreover, "predetermined chromosome patterns occur in some materials intensively studied." The chromosome patterns observed in AL can be genetically determined, resulting from interrelations between the genotype of the host and the malignant process itself. A possible example of the genetic influence of the host has been shown by Sandberg et al. (1966). These authors studied a pair of dizygotic twins presenting acute myeloblastic leukemia: an identical leukemic process appeared in both at the same age (7 months). Indeed, it could be postulated that environmental factors had been the same for both twins since conception. However, one child had a hypoploid cell line with 45 chromosomes while the other had a hyperploid cell line with 52 chromosomes. Very likely, each child responded to the neoplastic agent following its own genome. Still, from a fundamental point of view, the existence of leukemias with apparently normal chromosomes is very interesting. Drastic genome changes are thus not necessary for the transformation of a normal cell into a malignant cell. If cellular genetic processes are linked with neoplasia, this could indicate that small alterations in the genome, which could even be point mutations, are sufficient to explain the presence of leukemic cells in at least 50 percent of all cases.

Finally, one cannot exclude that a specific chromosome abnormality exists for a given form of AL. A new approach to this problem offered by recent techniques, such as fluorescence or study of banding patterns of chromosomes, merits further research.

The present chapter shall be divided into four sections: acute myeloblastic (or granulocytic) leukemia (AGL), acute lymphocytic leukemia (ALL), acute monocytic leukemia (AML), and other forms of acute leukemia. The difficulties in nomenclature encountered in myeloproliferative disorders are found with AL; on the other hand, in many cases the only information available concerns chromosome numbers and not karyotypes (See p. 92).

ACUTE GRANULOCYTIC LEUKEMIAS (AGL)

A summary of 113 cases of AGL (Table XII) has been given by Sandberg *et al.* (1968). Such a large series has a double advantage: first, observation and interpretation of results is highly uniform with the use of the same criteria from case to case; second, all cases are reported, not only those presenting a chromosome abnormality (a tendency of reports which concern but a single or a few cases). Briefly characterized, 50 percent of all cases of AGL have a normal karyotype, 21.2 percent present hypoploid stemlines. These characteristics have been found by others (Table XIII). Since a special section is devoted to leukemias in patients with a congenital chromosome abnormality, these cases have not been included in Table XIII. The presence of hypoploid cell lines is not usual in human cancer and thus is worthy of mention. For example, they are not found in acute lymphoblastic leukemia (*vide infra*); however, a higher frequency of hypoploid cell lines has been reported in erythroleukemia.

TABLE XII
SANDBERG'S FINDINGS IN 113 CASES OF AGL

No. of Chromosomes	41-45	46 (pseudo)	46	47-50	51-55
No. of Cases	24	8	57	20	4
%	21.2	7.0	50.0	17.7	4.1

TABLE XIII
KARYOTYPES OF PATIENTS WITH AGL

Author	Hypo	Pseudo	Diplo	Hyper	Total No. Cases
Baguena-Candela et al. (1964)	—	—	—	1	1
Baikie et al. (1961)	—	—	4	2	6
Bayreuther (1960)	—	—	2	—	2
Berger (1963)	—	—	2	—	2
Bottura et al. (1963)	—	—	1	—	1
De Nava (1969)	—	2	10	4	16
Elves et al. (1963)	—	—	—	1	1
Fortune et al. (1962)	—	1	—	—	1
Gavosto et al. (1963)	—	—	—	2	2
Hungerford and Nowell (1962)	1	—	5	3	9
Hossfeld et al. (1971)	—	2	—	—	2
Kahn et al. (1967)	—	—	—	1	1
Kinlough (1961)	—	—	4	2	6
Kiossouglou et al. (1965)	1	5	10	7	23
Lampert (1968)	—	—	5	—	5
Nowell and Hungerford (1960)	—	—	1	1	2
Pedersen (1964)	—	—	—	1	1
Reisman et al. (1964)	2	—	—	2	4
Ruffié et al. (1962)	3	—	7	—	10
Ruffié et al. (1964)	—	1	—	—	1
Sandberg et al. (1968)	24	8	57	24	113
Schleiermacher et al. (1967)	—	1	—	—	1
Stahl et al. (1966)	2	—	13	—	1
Weinstein and Weinstein (1963)	—	—	—	1	1
Whang-Peng et al. (1970)	3	4	73	7	103
Total	36	24	214	59	329
%	10.9	6.2	65.0	17.9	100

Note: In Whang-Peng's report, only 14 of 30 aneuploid karyotypes were described, so the proportion of diploid cases is somewhat lower than the 65% figure. In some other reports, a normal cell line coexists with an aneuploid one; only the aneuploid has been reported in the table.

The reader should note the great proportion of cases with apparently normal chromosomes. In these cases cytogenetics is of no value in diagnosis, but as stated in the introduction, it has great fundamental importance. This point will be discussed later. Several cases with a Ph^1 chromosome have been reported (Baguena-Candela and Fortezza Bover, 1964; Fortune et al., 1962; Hossfeld et al., 1971; Kiossouglou et al., 1965; Mastrangelo et al., 1967; Whang-Peng et al., 1970). It is not always possible to ascertain whether observation of the patient was made during blastic crisis of CML, as reflected, for example, by the observation of Mastrangelo et al. (1967): "abortive CML preceded by a blastic phase indistinguishable in all respects from AGL." In some cases, however, the nature of AGL with the Ph^1 chromosome and without CML seems established (Hosfeld et al., 1971): these cases are interesting because they are apparent exceptions to the specificity of the Ph^1 chromosome. Whang-Peng et al. (1970) proposed to isolate AGL with Ph^1 as a separate entitl. Ruffié et al. (1962) observed three cases with absence of a G-chromosome. Later, Lejeune (1965) asked whether this was not "total deletion," the Ph^1 being "partial deletion" of a G-chromosome. While partial deletion should lead to the chronic disease, total deletion should be linked with an acute form of disease. This statement, however, is difficult to prove because cases with a 45, G-karyotype are not sufficiently frequent and because this hypothesis was based on the disputed fact that "leukopoietic genes" are present on a G-chromosome.

ACUTE LYMPHOBLASTIC LEUKEMIA

If one examines the large series of cases of acute lymphoblastic leukemia (Sandberg et al., 1968 in Table XIV), it may be seen that there is no example of hypoploidy, that more than 50 percent of all cases have a normal karyotype, and that ALL is characterized chiefly by hyperploidy reaching tetraploidy. These results are in agreement with observations of various authors on different cases (Table XV). Only a few are hypoploid, the vast majority being hyperploid (50 to 59 in the Lampert series, 48 to 61 chromosomes in the Reisman series). Again, more than 55 percent of cases (including, however, some studied only from peripheral blood) have a normal karyotype.

ACUTE MYELOMONOCYTIC LEUKEMIA

A series of nine patients presenting microchromosomes was reported by Pierre et al. (1971). These microchromosomes were found together with other aneuploidies. Their number varied from 1 to 6. If each microchromosome was counted as a chromosome, hyperploid cell lines showed the loss of two or three "large chromosomes," being compensated for by the presence of two to four microchromosomes (see Fig. 13).

The presence of microchromosomes is rather frequent in tumors of the nervous system (see Chap. XVI), but uncommon in leukemias.

TABLE XIV
SANDBERG'S SERIES OF 106 PATIENTS WITH ALL

No. of Chromosomes	41-45	46*	47-50	51-55	58-62	90	Total
No. of Cases	0	46	21	16	8	6	106
%	0	50.9	19.8	15.0	7.5	6.8	100

*46 pseudodipl.

TABLE XV
ACUTE LYMPHOBLASTIC LEUKEMIA

Authors	Hypo	Pseudo	Diplo	Hyper	Total	Remarks
Baikie et al. (1961)	—	—	2	—	2	Blood cultures
Bayreuther (1960)	—	—	3	—	3	
Berger (1963)	—	1	3	—	3	
Chitham and McIver (1964)	—	—	—	—	1	
De Nava (1969)	—	—	8	—	8	
Fialkow et al. (1966)	—	1	1	—	1	
de Grouchy and Lamy (1962)	—	—	4	—	1	
Hungerford et al. (1961)	2 (C-1)	1	4	—	4	
Kiossoglou et al. (1965)	—	—	6	—	7	
Lampert (1968)	—	—	—	5	11	
Pearson et al. (1963)	—	—	—	2	2	Monozygotic twins
Ruffié et al. (1966)	—	1	—	1	1	
Sandberg et al. (1968)	1	—	54	51	106	
Stahl et al. (1966)	—	—	—	—	1	
Whang-Peng et al. (1969)	2	2	24	9	37	
Total	5	6	109	68	188	
%	2.8	3.2	57.9	36.1		

Note: No constant abnormality has been observed; the gains of chromosomes involved all groups.

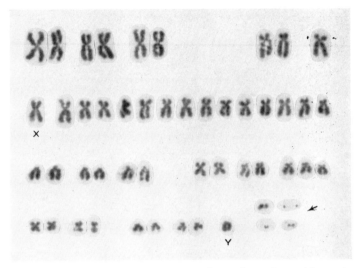

Figure 13. Karyotype of the bone marrow of a male patient with myelomonocytic leukemia showing four sets of double microchromosomes. On repeat examinations they were seen in almost all metaphases in bone marrow and only a small portion of metaphases from peripheral leukocytes. (Courtesy of Dr. R. V. Pierre et al., Mayo Clinic, Cancer, 27:160, 1971.)

Todd et al. (1969) observed another case of AML with microchromosomes. Rearrangements observed in pseudodiploid and hyperploid cell lines include translocations and, in one instance, rings (Baikie et al. 1959).

Stemlines with apparently normal chromosomes are slightly less frequent in this form of acute leukemia than in the other two already discussed (33% vs. 50%), and microchromosomes cannot yet be considered to be specific (see Table XVI).

OTHER FORMS OF ACUTE LEUKEMIA

Some cases of acute leukemia cannot be classified in the above three categories (AGL, ALL, and AML) either because another term has been employed in naming it, or because the case has been reported as acute leukemia without further classification (Table XVII). Some may be considered as variants of AGL, as indicated by the authors themselves (Fitzgerald et al., 1964; Knospe and Gregory, 1970) including a case of benzene leukemia (Forni and

TABLE XVI
CHROMOSOME FINDINGS IN 31 CASES OF MYELOMONOCYTIC LEUKEMIA

Author	Hypo	Pseudo	Diplo	Hyper	Total
Atkin and Goulian (1965)	—	—	—	1 (G+)	1
Baikie et al. (1961)	—	1	1	1	3
Baikie et al. (1959)	—	1	—	1 (r)	1
Berger (1963)	—	—	1	—	1
De Nava (1969)	—	—	1	1	2
Hungerford and Nowell (1962)	—	2	5	1	8
Kiossoglou et al. (1965)	—	—	3	—	3
Nowell and Hungerford (1962)	—	—	—	—	—
Pierre et al. (1971)	4	1	—	4	9 (all with microchr.)
Ruffié et al. (1964)	1	—	—	—	1
Todd et al. (1969)	1	—	—	—	1
Total	6	5	10	10	31

TABLE XVII
REPORTS OF ACUTE LEUKEMIAS WHICH CANNOT BE CLASSIFIED IN A DEFINITE GROUP

Author	Disorder	Hypo	Pseudo	Diplo	Hyper	Total
Borges et al. (1962)	AL in remission	—	—	22	—	22
De Nava (1969)	Acute promyelocytic L.	—	—	2	1	3
	Acute monoblastic L.	—	—	1	—	1
Engel (1965)	Aleukemic leukemia	1 (Ph¹, C—)	—	—	—	1
Fitzgerald et al. (1964)	AL (mostly AGL)	6	—	—	5 (3 polypl.)	11*
Forni and Moreo (1967)	Benzene leukemia	—	—	—	1	1
Goh et al. (1965)	Eosinophilic leukemia	—	—	2	—	2
Gunz et al. (1969)	Acute undifferentiated L.	—	1 (Ph¹)	—	—	1
Hammouda (1963)	AL	—	—	—	1	1
Hungerford (1961)	AL	—	—	—	—	7
Kemp et al. (1961)	AL	—	—	—	1 (C+)	1
Kiossouglou et al. (1965)	Promyelocytic L.	1	—	—	1	2
Knospe and Gregory (1971)	Smoldering acute lekemia	—	—	6	—	6
Pergoraro et al. (1963)	Agranulocytic leukemia	2	—	—	—	2
	AL	1	1	—	—	1
Pileri et al. (1966)	Promyelocytic L.	—	—	—	—	1
Reisman et al. (1964)	Acute stem cell leukemia	—	—	1	10	11
Total		11	2	41	20	74
%		16.8	2.8	55.4	27.0	

*Seven other cases only studied at the blood level, with normal chromosomes.

Moreo, 1967); other cases may be interpreted as variants of CML (Engel, 1965: Ph^1 positive aleukemic leukemia; Goh *et al.*, 1965: eosinophilic leukemia). It is in these forms of AL that hypodiploid counts are found, which is a general property of AGL. It must also be stressed that among this heterogenous group of acute leukemias, there are still more than 50 percent of cases that have ostensibly normal chromosomes. Indeed, in the series of Hungerford (1961), only chromosomes at blood level have been studied, possibly explaining why his findings were normal. As demonstrated by Fitzgerald *et al.* (1964), seven of 18 cases of AL studied only in blood were found to be normal. This contrasted sharply with 11 cases in which bone marrow was examined, all showing aneuploidies. Nevertheless, even excluding these cases, about 50 percent of cases having normal karyotypes remain. The case of Gunz *et al.* (1969) is of special interest since this represents a rare example of changing karyotype in AL, a feature more commonly observed in CML.

COMMENTS

The widespread nature of chromosome abnormalities in AL, together with the occurrence of 50 percent of those having normal karyotypes, makes it difficult to accept the hypothesis that chromosome aberration either is the cause, or that it is consistently correlated with the cause of the disease. This situation contrasts with the specificity of the Ph^1 chromosome in CML.

The clinical limits of application of cytogenetics in AL are evident. However, although still difficult to interpret, chromosome studies in AL are interesting since they make for working hypotheses concerning the possible causes of leukemia at the genome level. The first hypothesis is that if chromosomes are actually normal in more than 50 percent of cases, possibly an important alteration of the genome is not necessary for transformation of a normal cell into a leukemic cell. Even point mutations could be sufficient to account for malignancy; other chromosome abnormalities, when they exist, would thus be epiphenomena of the disease (Sandberg, 1966). When tissues other than bone marrow have been examined, they have been normal even if an aneuploid cell line was present in the hematopoietic tissue. This could mean that if a point mutation is sufficient to induce a leukemic process, it occurs at the cellular level and seems not to be inherited. The mean survival period is not different for patients with or without an aneuploid cell line, and depends upon the form of the disease.

A remarkable fact in AL is the stability of aneuploid cell lines in any given patient: when aneuploidy exists, it definitely seems linked with the propagation of the leukemia. This persistency of aneuploidy in AL contrasts with the instability of aneuploid cell lines in CML.

The simplest interpretation of AGL with a Ph^1 chromosome is that we actually are viewing blastic crisis of an undetected CML. Still, this explanation is not completely satisfying, and the meaning of this aneuploidy might include the possibility of a rearrangement such as translocation or deletion that could occur in any leukemia; the deleted G-chromosome could then belong to the pair not involved in formation of the Ph^1 chromosome.

Further work is needed to detect a specific aberration for a given form of AL. At present, this possibility still cannot be excluded.

CONCLUSIONS

1. No specific chromosome abnormalities have been found in any form of acute leukemia. Moreover, 55 percent of reported cases have apparently normal karyotypes.

2. Identical clinical forms of AL may show different chromosome patterns: each chromosomal anomaly appears to be unique. However, when an abnormal karyotype is encountered, it exhibits great stability. Acute granulocytic leukemias manifest more hypoploid cell lines, acute lymphoblastic leukemias more hyperploid lines.

3. Karyotype analysis alone cannot be used in AL either for diagnosis or for prognosis concerning survival. However, the presence of an aneuploid cell line allows the investigation and follow up of the progression of the disease.

4. Interpretation of the data is still difficult, but offers a basis for better understanding of basic concepts concerning causal effects of leukemia. Very small alterations of the genome, such as point mutations, might account in some cases for transformation of a normal cell into a leukemic cell.

REFERENCES

Atkins, L., and Goulian, M.: Multiple clones with increase in number of chromosomes in the G group in a case of myelomonocytic leukemia. *Cytogenetics, 4*:321, 1965.

Baguena-Candela, R., and Forteza-Bover, G.: Estudio citogenetico de una leucemia aguda del tipo mieloblastico promelocitico con cromosoma Filadelfia (Ph[1]). *Med Esp, 50*:298, 1964.

Baikie, A. G.; Court Brown, W. M.; Jacobs, P. A.; Milne, J. S.: Chromosome studies in human leukemia. *Lancet, 2*:425, 1959.

Baikie, A. G.; Jacobs, P. A.; McBride, J. A.; Tough, I. M.; Cytogenetic studies in acute leukemia. *Br Med J, 1*:1564, 1961.

Bayreuther, K.: Chromosomes in primary neoplastic growth. *Nature, 186*:6, 1960.

Berger, R.: Contribution a l'étude cytogenetique des leucemies humaines. Theses de Medicine, Paris, 1964.

Borges, W. H.; Nicklas, J. W.; Hamm, C. W.: Prezygotic determinants in acute leukemia. *J. Pediatr, 70*:180, 1967.

Bottura, C., and Ferrari, I.: Endoreduplication in acute leukemia. *Blood, 21*:207, 1963.

Chitham, R. G., and McIver, E. J.: Chromosome abnormality with lymphoid leukemia. *Lancet, 1*:1044, 1964.

De Nava, C.: Les anomalies chromosomiques au cours des hemopathies malignes et non malignes. *Monogr Ann Genet* (Paris), l'Expansion ed., 1969.

Engel, E.: Chromosomes in aleukemic leukemia. *Lancet, 2*: 1242, 1965.

Fitzgerald, P. H.; Adams, A.; Gunz, F. W.: Chromosome studies in adult acute leukemia. *J Nat Cancer Inst, 32*:395, 1964.

Forni, A., and Moreo, L.: Cytogenetic studies in a case of benzene leukemia. *Europ J Cancer, 3*:251, 1967.

Fortune, D. W.; Lewis, F. J. W.; Poulding, R. H.: Chromosome pattern in myeloid leukemia in a child. *Lancet, 1*:537, 1962.

Gavosto, F.; Pergoraro, L.; Pileri, A.: Possibilité de marquer à l'aide de précurseurs tritiés les chromosomes de cellules leucémiques chez l'homme. *Rev Franc Etud Clin Biol, 8*:920, 1963.

Goh, K.; Swisher, S. N.; Rosenberg, C. A.: Cytogenetic studies in eosinophilic leukemia. The relationship of eosinophilic leukemia and chronic myelocytic leukemia. *Ann Int Med, 62*:80, 1965.

de Grouchy, J., and Lamy, M.: Délétion partielle d'un chromosome moyen dans une leucémie aiguë lymphoblastique. *Rev Franc Etud Clin Biol, 7*:639, 1962.

Gunz, F. W.; Ravich, R. B. M.; Vincent, P. C.; Stewart, J. H.; Crossen, P. E.; Mellor, J.: A case of acute leukemia with a rapidly changing chromosome constitution. *Ann Genet, 13*:79, 1970.

Hammouda, F.: Chromosome abnormality in acute leukemia. *Lancet, 2*:410, 1963.

Hossfeld, D. K.; Han, T.; Holdsworth, R. N.; Sandberg, A. A.: Chromosomes and causation of the human cancer and leukemia. VII. The significance of the Ph[1] in conditions other than CML. Cancer, 27: 186, 1971.
Hungerford, D. A.: Chromosome studies in human leukemia. I. Acute leukemia in children. J. Nat Cancer Inst, 27:983, 1961.
Hungerford, D. A., and Nowell, P. C.: Chromosome studies in human leukemia. III. Acute granulocytic leukemia. J. Nat Cancer Inst, 29:545, 1962.
Kahn, M. H., and Martin, H.: G21 trisomy in a case of acute myeloblastic leukemia. Acta Haemat (Basel), 38:142, 1967.
Kemp, N. H.; Stafford, J. C.; Tanner, R. K.: Acute leukemia and Klinefelter's syndrome. Lancet, 2:434, 1961.
Kinlough, M. A., and Robson, H. N.: Study of chromosomes in human leukemia by a direct method. Br Med J, 2:1052, 1961.
Kiossoglou, K. A.; Mitus, W. J.; Dameshek, W.; Chromosomal aberrations in acute leukemia. Blood, 26:610, 1965.
Knospe, W. H.; Gregory, S. D.: Smoldering acute leukemia. Clinical and cytogenetic studies in six patients. Arch Intern Med, 127:910, 1971.
Lampert, F.: Kerntrockengewicht, DNS-Gehalt und Chromosomen bei akuten Leukämien im Kindesalter. Virchow Arch Abt Zellpath, 1:31, 1968.
Lejeune, J.: Leucémie et cancer. In Turpin, R. and Lejeune, J.: Les Chromosomes Humains. Gauthiers-Villars ed., Paris 1965, p. 200.
Levan, A.: Nonrandom representation of chromosome types in human tumor stemlines. Hereditas, 55:28, 1966.
Levan, A.: Chromosomes and carcinogenesis. In Handbook of Molecular Cytology. A. Lima de-Faria, (ed.), North-Holland Publishing Company, Amsterdam and London, 1969, p. 716.
Mastrangelo, R.; Zuelzer, W. W.; Thompson, R. I.: The significance of the Ph[1] chromosome in acute myeloblastic leukemia: serial cytogenetic studies in a critical case. Pediatrics, 40:834, 1967.
Muldal, S.; Elejalder, R.; Harvey, P. W.: Specific chromosome anomaly associated with autonomous and cancerous development in man. Nature, 229:48, 1971.
Nowell, P. C., and Hungerford, D. A.: Chromosome studies on normal and leukemic human leukocytes. J Nat Cancer Inst, 25:85, 1960.
Pearson, H. A.; Grello, F. W.; Cone, T. E.: Leukemia in identical twins. N Engl J Med, 268:1151, 1963.
Pedersen, B.: Two cases of chronic myeloid leukemia with presumably identical 47 chromosome cell lines in the blood. Acta Path Microbiol Scand, 61:497, 1964.
Pierre, R. V.; Hoagland, H. C.; Linman, J. W.: Microchromosomes in human preleukemia and leukemia. Cancer, 27:160, 1971.
Pileri, A.; Pergoraro, L.; Gavosto, F.: Cytogenetical and proliferative characteristics of acute promyelocytic leukemia cells. J Europ Cancer, 2:189, 1966.

Reisman, L. E.; Zuelzer, W. W.; Thompson, R. I.: Further observations on the role of aneuploidy in acute leukemia. *Cancer Res, 24*:1448, 1964.
Ruffié, J., and Lejeune, J.: Deux cas de leucémie aigue myéloblastique avec cellules sanguines normales et cellules haplo (21 ou 22). *Rev Franc Etud Clin Biol, 7*:644, 1962.
Ruffié, J.; Bierme, R.; Ducos, J.; Colombiès, P.; Salles-Mourlan, A. M.; Sendrial, A.: Haplosomie 21 associe a une trisomie 6-12 chez un tuberculeux presentant une myeloblastose sanguine et medullaire. *Nouv Rev Fran Hemat, 4*:719, 1964.
Ruffié, J.; Colombiès, P.; Combes, P. F.; Ducos, J.: Leucémie lymphoblastique chez un sujet porteur d'une anomalie congénitale complexe (type XXY probable). *Bull Acad Nat Med, 150*:342, 1966.
Sandberg, A. A.: The chromosomes and the causation of human cancer and leukemia. *Cancer Res, 26*:2064, 1966.
Sandberg, A. A.; Cortner, J.; Takagi, N.; Moghadam, M. A.; Crosswhite, L. H.: Differences in chromosome constitution of twins with acute leukemia. *N Engl J Med, 275*:809, 1966.
Sandberg, A. A.; Takagi, N.; Sofuni, T.; Crosswhite, L. H.: Chromosomes and causation of human cancer and leukemia. V. Karyotypic aspects of acute leukemia. *Cancer, 22*:1268, 1968.
Schleiermacher, E.; Kroll, W.; Hertl, M.; Heinhardt, N.: A constant chromosome aberration in two children with acute myeloid leukemia. *Humangenetik, 5*:80, 1967.
Stahl, A.; Papy-Jurion, M. C.; Muratore, R.; Mongin, M.; Olmer, J.: Estudio de los cromosomas en 35 observationes de leucosis. *Presse Med, 74*:429, 1966.
Todd, A. S.; Wood, S. M.; Robertson, J.; Brown, R. A. G.: A case of leukemia showing mixed myeloid-lymphoid characteristics and an unusual chromosome pattern. *J Clin Path, 22*:743, 1969.
Van Steenis, H.: Chromosomes and cancer. *Nature, 209*:819, 1966.
Weinstein, A. N., and Weinstein, E. D.: A chromosomal abnormality in acute myeloblastic leukemia. *N Engl J Med, 268*:253, 1963.
Whang-Peng, J.; Henderson, E. S.; Knutsen, T.; Freireich, E. J.; Gart, J. J.: Cytogenetic studies in acute myelocytic leukemia with special emphasis on the occurrence of Ph[1] chromosome. *Blood, 36*:448, 1970.
Whang-Peng, J.; Freireich, E. J.; Oppenheim, J. J.; Frei, E.; Tjio, J. H.: Cytogenetic studies in 45 patients with acute lymphocytic leukemia. *J Nat Cancer Inst, 42*:881, 1969.

Chapter IX

MYELOPROLIFERATIVE DISORDERS

According to Rappaport (1966), the myeloproliferative diseases (MD) are characterized by sustained or progressive proliferation of cells of the granulocytic, megakaryocytic, or erythrocytic lines, or of any combination of these three types of bone marrow cells. The relationship among these hematological disorders is shown in Table XVIII.

Some MD are not really neoplastic in nature. Even CML could be considered as benign, the Ph^1 chromosome representing a chromosomal instability predisposing to malignant transformation, in which case it would be the cytogenetic abnormality having the highest risk for acute leukemia (Zuelzer and Cox, 1969). Since one given myeloproliferative disorder can transform into another (see Table XVIII),

TABLE XVIII
RELATIONSHIP BETWEEN THE
DIFFERENT MYELOPROLIFERATIVE DISORDERS

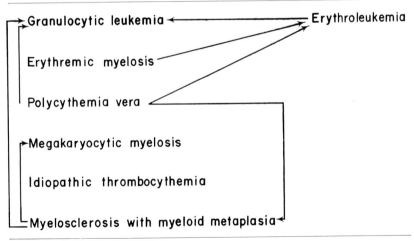

Note: From Rappaport, 1966.

clinical distinction between them is not always easy; transitional forms very often prevent us from distinguishing clearcut entities. However, since CML is involved in this group, the role of cytogenetics is obvious: the specificity of the Ph^1 chromosome is sufficiently great to separate CML from other MD. Moreover, "the common view that the Ph^1 chromosome confers a selective advantage on the abnormal clone implies that it is a direct link in the causal chain of CML. If so, the clone should precede the development of the disease" (Zuelzer and Cox, 1969). Arguments supporting this opinion have already been presented in Chapter VII; other data will be considered here.

It seems very useful for clinicians to know whether a specific MD will subsequently transform into CML. For example, thrombocythemia may represent the first step toward polycythemia vera (PV) or CML (Bousser et al., 1970). Evolution of the disease in the sense of a leukemic process could thus be predicted earlier if the Ph^1 chromosome could be demonstrated. Another example is myelofibrosis: in 19 percent of untreated cases of CML, Tenzer et al. (1967) found myelofibrosis. It is also well known that therapeutic agents may be responsible for myelofibrosis.

The presence or absence of the Ph^1 chromosome may help in the diagnosis of leukemoid reactions observed in severe infectious processes, a classic but now rare example being miliary tuberculosis. Detection of the Ph^1 chromosome is not the only goal of cytogenetic studies in MD. Aneuploidies are often observed in acute leukemias (AL), and the presence of such anomalies in a given myeloproliferative disorder could indicate its transformation into AL. Finally, the presence of the Ph^1 chromosome only in CML indicates that the basic differences between the myeloproliferative disorders and chronic granulocytic leukemia are greater than their morphologic similarities would tend to suggest.

Analysis of cytogenetic studies in MD is not simple, owing to difficulties in classification. Discussion will be presented in three sections: polycythemia vera, erythroleukemia, and other myeloproliferative diseases.

POLYCYTHEMIA VERA

Cytogenetic data on more than 80 patients is available. Several reports (Table XIX) present only a few cases (Wahrman et al., 1962;

TABLE XIX
CYTOGENETIC REPORTS IN POLYCYTHEMIA VERA

Authors	No. of Cases	Hypo	Diplo	Pseudo	Hyper	Remarks
De Nava (1969)	1	—	1	—	—	Transformed into AL
Erkman et al. (1967)	43	1	—	—	—	Fp— in 6 cases
Kay et al. (1966)	16	4	27	10	2	Pseudo is Ph¹+
Kemp et al. (1961)	9	—	15	1	—	Pseudo are Ph¹+
Koulischer et al. (1967)	1	1	6	2	1	"Ph¹-like" chromosome
Levan et al. (1964)	2	—	—	2	—	
Levin et al. (1967)	3	—	1	1	1	
Nowell and Hungerford (1962)	9	2	5	1	2	
Nowell (1971)	1	—	—	—	1	Transformed into AL
Wahrman et al. (1962)						
Total	86	8	55	17	7	

Nowell and Hungerford, 1962; Levan et al., 1964; Kemp et al., 1964; Erkman et al., 1967; De Nava, 1969). Larger series include nine cases of Koulischer et al. (1967); 16 cases of Kemp et al. (1961); 43 cases of Kay et al. (1966); and nine cases of Nowell (1971). The number of patients studied by Hirschhorn and Bloch-Stachter (1970) has not been specified. Chromosome findings are not uniform for two main reasons: some cases were already treated by ^{32}P when cytogenetic analysis was performed, or the disease had transformed into AL in others. It is known that ^{32}P may be responsible for chromosome alterations (McDiarmid, 1965). The findings are presented in the following four groups.

PV with Normal Chromosomes

This is not an infrequent finding especially when the disease is still untreated: six of nine cases of Koulischer et al. (1967), 15 of 16 cases of Kemp et al. (1961), 27 of 43 cases of Kay et al. (1966), five of nine cases of Nowell (1971). In summary (Table XIX), 55 of 86 cases (65%) had normal chromosomes.

PV with a Possible Marker Fp-

Kay et al. (1966) found six of 43 cases having a small deletion of the short arm of chromosome F. It has been also observed in PV by Hirschhorn and Bloch-Stachter (1970). However, until further confirmation, this deleted chromosome cannot be considered to be a specific marker for PV, in the sense of the Ph^1 chromosome in CML. De Nava (1969) observed a similar marker in acquired idiopathic sideroblastic anemia, and could not find a common link between the disease she studied and PV, except sideroblastosis or a "malignant tendency."

PV with the Ph^1 Chromosome

The finding of the Ph^1 chromosome in a myeloproliferative disorder suggests its transformation (actual or expected) into CML. An example of Ph^1 positive PV, reported by Kemp et al. in 1961 and transformed into CML in 1964, has already been commented upon in the chapter concerning CML. Koulischer et al. (1967) reported two

Ph^1 positive PV cases. One was a 71-year-old patient in whom PV had been diagnosed in 1960; he was treated with ^{32}P. In 1965, his disease transformed into CML; in 1967, 72 percent of bone marrow cells were Ph^1 positive. The second case reported was that of a 74-year-old patient having PV still untreated when the cytogenetic analysis was performed. A Ph^1 chromosome was observed in 10 percent of the mitoses analyzed. However, two years later, the patient had still not developed CML. Levin *et al.* (1967) observed a Ph^1 chromosome in the bone marrow of two brothers with PV. The level of LAP (Leucocyte Alkaline Phosphatase) was moderately increased.

Interpretation of these unusual results is difficult. It was not possible for the authors to decide whether the disease and the chromosomal aberrations in these two brothers were familial or fortuitous. It would certainly be interesting to understand this observation, to study the chromosomes of a tissue other than bone marrow, and to follow the evolution of the disease.

PV with Aneuploid Cells or Aneuploid Cell Clones

The finding of an aneuploid cell clone in PV is generally related to the transformation of the disease into AL (Nowell and Hungerford, 1962; Wahrman *et al.*, 1962; Nowell, 1965; Kay *et al.*, 1966; Erkman *et al.*, 1967; Koulischer et al., 1967). Because no specific abnormality has been observed, cytogenetic analysis cannot be used to distinguish AL developing in PV from other forms of acute leukemias. Non-specific anomalies such as fragments, breaks, rings, etc. were found chiefly in patients treated with ^{32}P.

ERYTHROLEUKEMIA

Dealing with data on erythroleukemia requires careful classification. This malignant myeloproliferative disorder has been reported in acute and chronic forms, under the terms of erythroleukemia, erythremic myelosis, or DiGuglielmo's syndrome. These are perhaps different diseases, but so similar to one another that distinction is at times impossible and thus they are grouped together in this paragraph. Some examples demonstrate the diversity of symptomatology and the difficulty in terminology. Di Grado *et al.* (1964) reported a case of DiGuglielmo's syndrome in which the clinical findings resembled those in erythroleukemia. Heath and Moloney (1965) studied a patient with "erythremic myelosis" considered as erythroleukemia by Crossen *et al.* (1969).

The preleukemic phase of the disease has been diagnosed as (a) megaloblastic anemia which persisted through the first four years of life before the leukemic nature of the process was recognized (McClure *et al.*, 1965), (b) a hypoplastic constitutional anemia (Weatherall and Walker, 1965), or (c) polycythemia vera (Kay *et al.*, 1966). Two cases were treated by x-radiation for ankylosing spondylitis (Buckton *et al.*, 1962). In one report, the terminal phase was described as acute leukemia with paramyeloblasts (Baserga and Ricci, 1964).

Despite the apparent diversity, some common characteristics can be observed. The summary of the results is shown in Table IV. Note that almost 40 percent of all cases showed hypoploidy. This general tendency towards hypoploidy is not frequent in human cancer.

No specific loss of a particular chromosome was reported (Table XX) and, in the same case, cells with 45 chromosomes showed different karyotypes (Crossen *et al.*, 1969).

TABLE XX
REPORTS OF 54 CASES OF ERYTHROLEUKEMIA

Authors	No. of Cases	Hypo	Diplo	Pseudo	Hyper/ Polypl.
Adachi *et al.* (1964)	1	—	—	—	1
Baserga and Ricci (1964)	1	—	—	1	—
Baikie *et al.* (1961)	2	—	1	—	1
Becak *et al.* (1967)	1	—	—	1	—
Buckton *et al.* (1962)	2	—	—	—	2
Castoldi *et al.* (1968)	6	2	1	2	1
Crossen *et al.* (1969)	1	1	—	—	—
De Nava (1969)	1	—	1	—	—
Di Grado *et al.* (1964)	1	1	—	—	—
Durant and Tassoni (1967)	2	1	—	—	1
Dyment *et al.* (1968)	3	1	2	—	—
Engel *et al.* (1967)	1	1	—	—	—
Fitzgerald *et al.* (1964)	1	—	—	—	1
Hayhoe and Hammouda (1965)	1	—	—	—	1
Heath and Moloney (1965)	1	1	—	—	—
Kay *et al.* (1966)	1	1	—	—	—
Kiossouglou *et al.* (1966)	16	5	7	2	2
Krogh-Jensen (1969)	6	2	4	—	—
Krompotic *et al.* (1968)	1	—	—	—	1
McClure *et al.* (1965)	1	1	—	—	—
Pawelski *et al.* (1963)	2	2	—	—	—
Stahl *et al.* (1965)	1	1	—	—	—
Weatherall and Walker (1965)	1	—	—	—	1
Total	54	20	16	6	12

Note: No specific loss of chromosomes was observed in hypoploid cell lines.

Polyploidy was also frequent and seems a basic feature of the disease (Crossen et al., 1969). Polyploid cells arise from stemline cells as demonstrated in Heath and Moloney's report; the stemline showed 39 chromosomes, and the polyploids 78 and 156. In the case of Crossen et al., there was a stemline with 45 chromosomes, a ring chromosome, and a number of polyploid cells with counts of 7n and 8n containing 4 ring chromosomes. In addition to this case, rings have been observed by Di Grado et al. (1964) and Krogh-Jensen (1969). The ring, however, was of different origin in these cases (groups A, B, C, and F) but probably might represent "a measure of the incidence of structurally abnormal cell lines in this disorder" (Crossen et al., 1969).

OTHER MYELOPROLIFERATIVE DISEASES

Cases grouped in this paragraph include *myeloid metaplasia* with or without myelofibrosis, *myelofibrosis* alone, *thrombocytosis* or thrombocythemia (essential or symptomatic), *myeloid splenomegaly*, which is a French term for *myelofibrosis with myeloid metaplasia*. Also all cases reported under the general category of myeloproliferative disorder ("usual" or "unusual"), including myeloid aplasia, atypical chronic granulocytic leukemia, and granulocytic hyperplasia. Needless to say, this diverse terminology reflects the clinical difficulties involved in diagnosis. An attempt at classification of the different MD, using the terminology of the authors, is proposed in this text. However, for verification of the clinical diagnosis, the reader should refer to the original reports.

Thrombocythemia (Thrombocytosis)

Cytogenetic analyses (Table XXI) show that among 13 cases, 4 had normal karyotype, three were Ph^1 positive cases, one developed CML (Woodliff et al., 1966), the second was also included in a series of CML (Tough et al., 1963) and, in the third, the diagnosis of CML was uncertain (Heath and Moloney, 1965). In these patients, the thrombocytosis was in close relationship to CML clinically and cytogenetically. The two patients with supernumerary C-chromosomes did not develop acute leukemia which suggests that the appearance of an aneuploid cell line in the bone marrow does not necessarily mean malignant transformation of the disease.

TABLE XXI
REPORTS OF 13 CASES OF THROMBOCYTHEMIA

Authors	No. of Cases	Karyotype	Remarks
De Nava (1969)	3	N	
Heath and Moloney (1965)	1	46, Ph¹	Atypic myeloproliferative syndrome
Kiossouglou et al. (1965)	3	N	
Koulischer (1969)	1	N	
Rowley et al. (1966)	1	48, 2C+	"Idiopathic thromb"
Tough et al. (1963)	1	46, Ph¹	
Woodliff et al. (1966)	3	46, Ph¹	evolution into CML
		47, C+	myelofibrosis?
		N	evolution into PV

TABLE XXII
DESCRIPTION OF THE DIFFERENT KARYOTYPES OBSERVED IN MYELOID METAPLASIA AND MYELOID METAPLASIA WITH MYELOFIBROSIS

Authors	No. of Cases	Karyotype	Remarks
Bowen and Lee (1963)	1 (MMM)	46, Ph¹	evolution into CML
De Nava (1969)	1 (MMM)	46, Dq—, C—, 16+	
	6 (MM+M?)	N	
		46, D—, 3+	
		4 with anomalies	mainly hypoploid cells
Forrester and Louro (1966)	1 (MM)	46, Ph¹	
Holden et al. (1970)	1 (MMM)	45, C—	5-year-old child; death occurred; spleen with MM
Jackson and Higgins (1967)	1 (MMM)	45, C—	Septicemia and death.
Kiossouglou et al. (1966)	2 (MMM)	47, C+	
	2 (MMM)	N	
Koulischer (1969)	3 (MM)	2 N	
		1 with cells with 2n=47	No cell line could be demonstrated
Nowell and Hungerford (1962)	2 (MM)	46, E—, C+	Evolution in subacute granulocytic leukemia
		aspecific an.	
Padeh et al. (1965)	5 (MM)	N	
Sandberg et al. (1962)	12 (MM)	N	
Sandberg et al. (1964)	1 (MM)	47, C+	Possible leukemia
Solari et al. (1962)	1 (MM)	46, Cq—	MM after treated PV

Note: MM: myeloid metaplasia; MMM: myeloid metaplasia with myelofibrosis; N: normal karyotype.

Myeloid Metaplasia with or without Myelofibrosis

The chromosomes of 39 cases have been investigated (Table XXII). These were myeloid metaplasia (MM), myeloid metaplasia with myelofibrosis (MMM), or myelofibrosis alone (M). Karyotypes were normal in 22 cases; four had pseudodiploid cell lines. Chromosomes of the C group were involved in three cases; two had a supernumerary C and two others a missing C chromosome. Finally two were Ph^1 positive and one transformed to CML (Bowen and Lee, 1963). The frequency of the involvement of C group chromosomes is remarkable, but by no means can this be considered to be specific. Holden et al. (1970) proposed to isolate a new clinical entity: myelofibrosis with a 45, C- karyotype. This seems premature, since other myeloproliferative disorders with different clinical symptomatology may have such a karyotype (see next paragraph, Freireich et al., 1964; Teasdale et al., 1970). Nevertheless, there is no doubt that cytogenetics may prove to be useful in isolating closely related, but still different diseases. In this respect, more refined techniques including fluorescence and other banding patterns of chromosomes seem promising. However, there is as yet no proof that the "missing" C is the same in different cases with 45, C- cell lines.

"Myeloproliferative Syndromes"

In some reports, the exact nature of the myeloproliferative syndrome has not been stated. Many of these cases (Table XXIII), characterized by an aneuploid cell line, transformed into acute or subacute leukemia with rapid development. This was particularly clear in Nowell's series and in the report of Teasdale et al. (1970). However, exceptions do exist: a case of Nowell (1965) had a 48, F+G+ cell line and no change in the sense of a transformation into acute leukemia was observed after 43 months; the three cases of Freireich et al. (1964) with a 45, C- cell line were not AL, as well as the patient of Winkelstein et al. (1966) with a 47, C+ line. The reverse is rarely true: one case of Nowell, an atypical CML, had normal chromosomes. The patient died, however, seven months after cytogenetic analysis as a result of sudden onset of blastic crisis.

TABLE XXIII
KARYOTYPES REPORTED IN SOME UNDETERMINED "MYELOPROLIFERATIVE SYNDROMES"

Authors	No. of Cases	Karyotype	Remarks
Freireich et al. (1964)	3	45, C—	Granulocytic hyperplasia
Houston et al. (1964)	1	46, Ph^1, ?— mar	CML with very slow development
Kiossouglou et al (1966)	4	2 with 47, C+ 1 with 46, Ph^1 1 with Normal	
Lawler et al. (1966)	1	47, C+	Medullar dysplasia
Nau and Hoagland (1971)	1	Normal	Persistent basophilia, granulocytic leukemia
Nowell (1965)	10	46, F—, minute + 45, t (1-2) 47, D+ 48, F+G+ 6 with Normal	Rapid death, subacute granulocytic leukemia No change after 43 months One atypical CML
Nowell and Hungerford (1962)	3	Normal	
Teasdale et al. (1970)	3	45, C—	3 children; death, acute myeloblastic leukemia
Winkelstein et al. (1966)	1	47, C+	

CONCLUSIONS

The bone marrow chromosomes of more than 200 patients with a myeloproliferative disorder other than CML have been studied. It seems justified to draw the following conclusions:

1. The diversity in the terminology used by the different authors reflects the difficulties of diagnosis because myeloproliferative disorders are closely related to each other. Cytogenetics appears as a new tool in the study of these diseases, with the prospect that more precise separation among them will be possible after karyotyping.

2. No specific chromosome abnormality has been demonstrated in any of the myeloproliferative diseases other than CML, although such claims exist. Nevertheless, a remarkable finding is the high frequency (almost 40%) of hypoploid cell lines in erythroleukemia, and the frequent involvement of C group chromosomes in aneuploid cell lines of MD.

3. With a few inconclusive exceptions, the presence of a Ph^1 chromosome is related to the transformation of the disease into CML.

4. The cases in which the presence of an aneuploid cell line signi-

fied the transformation of the disease to acute or subacute leukemia (always of the myeloblastic form) can not constitute the basis for a general rule. Some cases show aneuploidy without neoplastic transformation which confirms that aneuploidy predisposes to, but is not necessary for, malignant transformation.

5. Application of new, more refined techniques for identification of chromosomes involved in aneuploidy will be of substantial help.

REFERENCES

Baserga, A., and Ricci, N.: *Cytogenetic Studies in Erythroleukemia.* Stockholm, 10th Congress of Int Soc Haemat, 1964.

Becak, W.; Becak, M. L.; Saraiva, L. G.: Cromosoma acrocentrico gigante en un caso de sindrome de Di Guglielmo. *Sangre, 12*:65, 1967.

Bousser, J.; Bilski-Pasquier, G.; De Grouchy, J.; Guernet, M.; Zittoun, J.; Bernadou, A.; De Nava, C.; Guillerm, M.; Fretault, J.; Zittoun, R.: Confrontation des données de l'étude chromosomique avec le taux de phosphatase alcaline leucocytaire et de la vitamine B12 sérique dans les syndromes myélo-prolifératifs. *Nouv Rev Franc Hemat, 10*:75; 1970.

Bowen, P., and Lee, C. S. N.: Ph¹ chromosome in the diagnosis of chronic myeloid leukema. Report of a case with features simulating myelofibrosis. *Bull Johns Hopkins Hosp, 113*:1, 1963.

Buckton, K. E.; Jacobs, P. A.; Court Brown, W. M.; Doll, R.: A study of the chromosome damage persisting after X-ray therapy for ankylosing spondylitis. *Lancet, 2*:676, 1962.

Castoldi, G.; Yam, L. T.; Mitus, W. J.; Crobsy, W. H.: Chromosomal studies in erythroleukemia and chronic erythremic myelosis. *Blood, 31*:202, 1968.

Crossen, P. E.; Fitzgerald, P. H.; Menzies, R. C.; Brehaut, L. A.: Chromosomal abnormality, megaloblastosis and arrested DNA synthesis in erythroleukemia. *J Med Genet, 6*:95, 1969.

De Nava, M.: Les anomalies chromosomiques au cours des hémopathies malignes et non malignes. *Monogr Ann Genet* (Paris), l'Expansion éd., 1969, Vol. 1.

DiGrado, F.; Mendes, F. T.; Schroeder, E.: Ring chromosome in a case of DiGuglielmo syndrome. *Lancet, 2*:1243, 1964.

Durant, J. R.; Tassoni, E. M.: Coexistent DiGuglielmo's leukemia and Hodgkin's disease. A case report with cytogenetic studies. *Am J Med Sci, 254*:824, 1967.

Dyment, P. G.; Melnyk, J.; Brubaker, C.: A cytogenetic study of acute erythroleukemia in children. *Blood, 32*:997, 1968.

Engel, E.; McKee, L. C.; Bunting, K. W.: Chromosomes 17-18 in leukemias. *Lancet, 2*:42, 1967.

Erkman, B.; Hazlett, B.; Crookston, J. H.; Conen, P. E.: Hypodiploid chromosome pattern in acute leukemia following polycythemia vera. *Cancer,* 20:1318, 1967.

Fitzgerald, P. H.; Adams, A.; Gunz, F. W.: Chromosome studies in acute leukemia. *J Nat Cancer Inst, 32*:395, 1964.

Forrester, R. H.; Louro, J. M.: Philadelphia chromosome abnormality in agnogenic myeloid metaplasia. *Ann Int Med, 64*:622, 1966.

Freireich, E. J.; Whang, J.; Tjio, J. H.; Levin, R. H.; Brittin, G. M.; Frei, E. III: Refractory anemia, granulocytic hyperplasia of bone marrow and a missing chromosome in marrow cells. A new clinical syndrome? *Clin Res, 12*:284, 1964.

Hayhoe, F. G. J., and Hammouda, F.: Cytogenetic and metabolic observations in leukemias and allied states. In F. G. J. Hayhoe (Ed.): *Current Research in Leukemia.* Cambridge, University Press, 1965.

Heath, C. W., and Moloney, W. L.: Cytogenetic observations in a case of erythremic myelosis. *Cancer, 18*:1495, 1965.

Heath, C. W., and Moloney, W. C.: The Philadelphia chromosome in an unusual case of myeloproliferative disease. *Blood, 26*:471, 1965.

Hirschhorn, K.; Bloch-Stachter, N.: Transformation of genetically abnormal cells. In *Genetic Concepts and Neoplasia.* Baltimore, Williams and Wilkins, 1970, vol. 1, pp. 191-202.

Holden, J. et al.: Myelofibrosis with C monosomy of marrow elements in a child. *Am J Clin Path, 55*:573, 1971.

Houston, E. W.; Levin, W. C.; Ritzmann, S. E.: Untreated chronic myelocytic leukemia associated with an unusual chromosome pattern. *Ann Int Med, 61*: 696, 1964.

Jackson, J. F., and Higgins, L. C.: Group C monosomy in myelofibrosis with myeloid metaplasia. *Arch Intern Med, 119*:403, 1967.

Kay, H. E. M.; Lawler, S. D.; Millard, R. E.: The chromosomes in polycythemia vera. *Br J Hemat, 12*:507, 1966.

Kemp, N. H.; Stafford, J. L.; Tanner, R. K.: Cytogenetic studies of polycythemia vera. Vienna, 8th Congress Europ Soc Haemat, 1961, p. 92.

Kemp, N. H.; Stafford, J. L.; Tanner, R.: Chromosome studies during early and terminal chronic myeloid leukemia. *Br J Med, 1*:1010, 1964.

Kiossouglou, K. A.; Mitus, W. J.; Dameshek, W.: Cytogenetic studies in the chronic myeloproliferative syndrome. *Blood, 28*:241, 1966.

Koulischer, L.; Fruhling, J.; Henry, J.: Observations cytogénétiques dans la maladie de Vaquez. *Europ J Cancer, 3*:193, 1967.

Koulischer, L.: Contribution à l'étude des chromosomes dans les leucémies humaines. Thèse, Université Libre de Bruxelles, 1968.

Krogh-Jensen, M.: Chromosomal findings in two cases of acute erythroleukemia. *Acta Med Scand, 180*:245, 1966.

Krogh-Jensen, M.: *Chromosome Studies in Acute Leukemia.* Munksgaard, Copenhagen, 1969, vol. 1.

Krompotic, E.; Silberman, S.; Einhorn, M.; Uy, E. S.; Chernay, P. R.: Clonal evolution in DiGuglielmo syndrome. *Ann Genet, 11*:225, 1968.

Levan, A.; Nichols, W. W.; Hall, B.; Löw, B.; Nilsson, S. B.; Nordén, Å.: Mixture of Rh positive and Rh negative erythrocytes and chromosomal abnormalities in a case of polycythemia. *Hereditas, 52*:89, 1964.

Levin, W. C.; Houston, E. W.; Ritzmann, S. E.: Polycythemia vera with chromosome mosaicism. Report of a case in a 5 year-old boy. *Arch Intern Med, 115*:697, 1965.

Macdiarmid, W. D.: Chromosomal changes following treatment of polycythemia with radioactive phosphorus. *Quart J Med, 34*:133, 1965.

Nau, R. C., and Hoagland, H. C.: A myeloproliferative disorder manifested by persistent basophilia, granulocytic leukemia and erythroleukemic phases. *Cancer, 28*:662, 1971.

Nowell, P. C.: Prognostic value of marrow chromosome studies in human "preleukemia." *Arch Path, 80*:205, 1965.

Nowell, P. C.: Marrow chromosome studies in "preleukemia." *Cancer, 28*:513, 1971.

Nowell, P. C., and Hungerford, D. A.: Chromosome studies in human leukemia. IV. Myeloproliferative syndrome and other atypical myeloid disorders. *J Nat Cancer Inst, 29*:911, 1962.

Padeh, B.; Bianu, G.; Schaki, R.; Akstein, E.: Cytogenetic studies in proliferative disorders. *Israel J Med Sci, 1*:795, 1965.

Pawelski, S., Toploska, P.; Maj, S.: Chromosome abnormalities in DiGuglielmo syndrome. *X Cong Soc Europ Hemat* Strasbourg, 1963.

Rowley, J. D.; Blaisdell, R. K.; Jacobson, L. O.: Chromosome studies in preleukemia. I. Aneuploidy of group C chromosomes in 3 patients. *Blood, 27*:782, 1966.

Sandberg, A. A.; Ishihara, T.; Crosswhite, L. H.; Hauschka, T. S.: Comparison of chromosome constitution in chronic myelocytic leukemia and other myeloproliferative disorders. *Blood, 20*:393, 1962.

Sandberg, A. A.; Ishihara, T.; Crosswhite, L. H.: Group-C trisomy in myeloid metaplasia with possible leukemia. *Blood, 24*:716, 1964.

Solari, A. J.; Sverdlick, A. B.; Vaola, E. R.: Chromosome abnormality in myeloid metaplasia. *Lancet, 2*:613, 1962.

Stahl, A.; Papy, M. C.; Muratore, R.; Mongin, M.; Olmer, J.: Erythromyélose avec pseudo-syndrome de Pelger-Huet et anomalies chromosomiques complexes. *Nouv Rev Franc Hemat, 5*:879, 1965.

Tanzer, J.; Najean, Y., Jacquillat, C., Ripault, J.; Chome, J.: Fibrose médullaire et érythroblastose splénique dans la leucémie myéloïde chronique. *Nouv Rev Franc Hemat, 7*:801, 1967.

Teasdale, J. M.; Worth, A. J.; Corey, M. J.: A missing group C chromosome in the bone marrow cells of three children with myeloproliferative disease. *Cancer, 25*:1468, 1970.

Tough, I. M.; Jacobs, P. A.; Court Brown, W. M.; Baikie, A. G.; Williamson,

E. R. D.: Cytogenetic studies on bone-marrow in chronic myeloid leukemia. *Lancet, 1*:844, 1963.

Wahrman, J.; Schaap, T.; Robinson, E.: Manifold chromosome abnormalities in leukemia. *Lancet, 1*:1098, 1962.

Weatherall, D. J., and Walker, S.: Changes in the chromosome and haemoglobin patterns in a patient with erythroleukemia. *J Med Genet, 2*:212, 1965.

Winkelstein, A.; Sparkes, R. S.; Craddock, C. G.: Trisomy of group C in a myeloproliferative disorder, Report of case. *Blood, 27*:722, 1966.

Woodliff, H. J.; Dougan, L.; Onesti, P.: Cytogenetic studies in twins, one with chronic granulocytic leukemia. *Nature, 211*:533, 1966.

Zuelzer, W. W., and Cox, D. E.: Genetic aspects of leukemia. In Holland, J. F., *et al.* (Ed.): *Leukemia and Lymphoma.* New York and London, Grune and Stratton, 1969, pp. 5-25.

Chapter X

NON-LEUKEMIC HEMATOLOGICAL DISORDERS: PRELEUKEMIAS, ANEMIAS, DYSPROTEINEMIAS

As in the case of myeloproliferative disorders, some nonleukemic hematological diseases, at times, seem so closely related to leukemia or seem to be so prone to malignant transformation, that it has been tempting to study their chromosomes. This heterogenous group comprises a wide variety of refractory or unexplained anemias: aplastic anemia (Fanconi's anemia), Waldenström's macroglobulinemia, multiple myeloma, other "gammopathies" (Houston et al., 1967), and ill-defined preleukemic states or myelodysplasias. Myelodysplasias as defined by Rowley et al. (1966) accurately characterize some of the diseases concerned in this chapter as: "a group of disorders exhibiting morphologic evidence in the marrow and the blood of disturbed formation of any, or any combination, of the major cell lineages (erythroid, granulocytic, megakaryocytic), not characteristic of the aplastic, the myeloproliferative, or the leukemic state, yet displaying features often intermediate or transitional between these conditions." Nowell (1971) included myeloproliferative diseases in "preleukemias"; in the present chapter only his results concerning diseases other than myeloproliferative will be discussed.

The goal of cytogenetics is, as usual, to demonstrate either the presence of a marker specific for a given disease or to establish correlation between aneuploidy and malignant transformation. A specific marker was thought to have been found in Waldenström's macroglobulinemia and the term "W" chromosome was proposed (Benirschke et al., 1962). However, since a similar abnormality was observed in other monoclonal gammopathies, the term "MG chromosome" was suggested by Houston et al. (1967) as more accurate. Other claims for markers have been made and will be discussed later in the text.

For the sake of clarity, results will be grouped in three sections. The first deals with all patients in whom anemia has been the main reason for chromosome study; the second concerns Waldenström's macroglobulinemia, and the third multiple myeloma and other diseases with abnormal paraproteins.

ANEMIC STATES

Refractory anemia is so very often a premonitory sign of leukemia that it is referred to as *preleukemia*. In many cases reported in the chapters concerning leukemias, cytogenetic studies were first undertaken because of unexplained severe anemia. For example, two patients described by Whang-Peng et al. (1970) were first investigated for refractory anemia; subsequently they developed acute granulocytic leukemia and were included in the group of AGL. Some patients of Knospe and Gregory (1971) reported under the term *smoldering acute leukemia* (one synonym proposed by the authors being preleukemia) had, in fact, anemia. Similar examples are numerous. However, true anemias, apparently unrelated to leukemia, do exist, and it was thought that perhaps cytogenetics might help in distinguishing true anemia from preleukemia. Different forms of anemia have been investigated, including pernicious anemia, aplastic anemia, sideroachrestic anemia, sideroblastic anemia, folate and vitamin B_{12} deficiency anemia, and idiopathic acquired anemia. As already stated by Heath (1966) "the several reports are difficult to interpret and compare, both because clinical and cytogenetical findings are not always fully described and because the tissues studied and the cytogenetic methods used vary." However, an attempt at synthesis is presented in Table XXIV.

The vast majority (75%) of these cases exhibit normal karyotype. Occasionally (from 2% to 8%), in some cases, nonspecific abnormalities such as breaks, gaps, and centromere spreading may be observed in some mitoses. Adequate antianemic treatment is followed by a decrease of these anomalies (Heath, 1966). No specific markers have been observed, although De Nava (1969) reported six cases of "idopathic sideroblastic acquired anemia" with an Fp- chromosome. A similar chromosome has been observed in some cases of polycythemia vera by Kay et al. (1966) as already noted in Chapter IX. Patients with obvious chromosomal abnormalities have failed to develop leukemia after several years of observation (Rowley et al., 1966; Nowell, 1971).

TABLE XXIV
ANEMIC STATES CYTOGENETICALLY INVESTIGATED

Author	Name of the Disorder	No. of Cases	Hypo	Pseudo	Diplo	Hyper	Remarks
Astaldi et al. (1962)	pernicious anemia	1	—	—	1	—	
Court Brown et al. (1960)	megaloblastic anemia	7	—	—	7	—	
de la Chapelle and Gräsbeck (1963)	pernicious anemia	4	—	—	4	—	
De Nava (1969)	acquired idiopathic sideroblastic anemia	12	—	6	6	—	blood cultures, vit B$_{12}$ free medium.
Ford et al. (1968)	pernicious anemia	2	—	—	2	—	
	pregnancy anemia	3	—	—	3	—	
Fortezza Bover and Baguena Candela (1963)	pernicious anemia	1	1	—	—	—	
Heath (1966)	vitamin B$_{12}$ and folate deficiency	14	—	—	14	—	nonspecific anomalies in some mitoses.
Kiossoughlou et al. (1965)	pernicious anemia	3	2	—	1	—	"Premyeloid leukemia"
Leeksma et al. (1965)	acquired sideroblastic anemia	—	—	—	—	—	
Nowell (1971)	sideroblastic anemia, pancytopenia	3	—	—	3	—	2 patients died
		16	5	—	11	—	1 erytheroleukemia 1 megakaryocytic leukosis.
Rowley et al. (1966)	aplastic anemia, sideroachrestic anemia, idiopathic anemia	3	—	—	—	—	supernumerary C group chromosome in the 3 cases
Total		69	8	6	52	3	

Note: The name of the disorder as given by the authors is reported in the table.

In *Fanconi's anemia* (cases not included in the table), nonspecific abnormalities in a significant number of mitosis are very often found (Bloom et al., 1966). These are gaps, breaks, chromosome exchanges, and endoreduplications. However, the number of chromosomes is normal. This disease is transmitted as an autosomal recessive trait and shows a high frequency of leukemic transformation. It has been suggested that the chromosome abnormalities observed could be related to the malignant evolution. From a cytogenetic point of view, Bloom's disease is very similar to Fanconi's anemia, also showing nonspecific karyotype anomalies and a high incidence of leukemia (German et al., 1965).

In fact, among the group of anemias presented in this section, chromosomes are usually normal and eventual aneuploidy is seldom linked with malignant transformation. Perhaps this is because the diseases reported are "true anemias." As an example, Pierre and Hoagland (1971) described the loss of the Y-chromosome in bone marrow of old anemic males. Their finding was not linked with leukemia since the same phenomenon was observed to be associated with

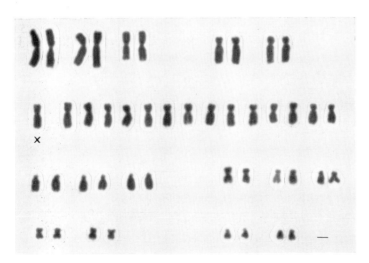

Figure 14. Karyotype from the bone marrow of a male patient, asymptomatic for pernicious anemia for four months (pernicious anemia was the original diagnosis). Out of 50 metaphases from bone marrow, 24 were lacking the Y chromosome. All metaphases (25) from peripheral blood were 46,/XY. (Courtesy of Dr. R. V. Pierre, Mayo Clinic, *Mayo Clin Proc, 46*:52, 1971.)

advanced aging (See Fig. 14). When anemia is a symptom of leukemia, in general, the leukemic process develops within a few months. Patients are then considered leukemic and reported in leukemic series. The significance of chromosome anomalies in true anemia is still not clear; they are far from being linked with leukemia.

WALDENSTRÖM'S MACROGLOBULINEMIA

This blood disease is characterized by the presence of a serum macroglobulin and is included in the group of the gammopathies. The abnormal protein is a gamma M; the disease frequently progresses towards malignancy. The initial report by Bottura et al. (1961) mentioned the presence of a clone with 47 chromosomes. The supernumerary chromosome was the size of a group A chromosome: 44 of the 90 mitoses analyzed (50%) showed the abnormality. This finding was soon confirmed by German et al. (1961) and by Benirschke et al. (1962) who introduced the term "W" chromosome. Since then, other cases have been reported (Table XXV) with or without the marker chromosome.

The finding of the marker in different cases deserves some comments. First, the morphology of the marker is not constant. In general, it is a large chromosome, the size of those of the A, B, or C groups. The position of the centromere may vary from subterminal to metacentric. In one case, the supernumerary chromosome was the size of an F-chromosome (Spengler et al., 1966).

Second, the marker, when present, is found in a variable proportion of mitoses (from 1% to 51% of all mitoses counted). The low proportion of abnormal cells could be explained by the fact that they divide more slowly than the normal cells. Since the marker has been observed in the marrow and peripheral blood, for the demonstration of the presence of the marker we recommend study of both tissues, counting as many cells as possible.

From a theoretical point of view, it is tempting to think that aneuploid lymphocytes are responsible for the secretion of the abnormal protein; even a small proportion of these could account for the presence of the macroglobulin. However, the relatively large series of negative cases of De Nava (1969), as well as other negative reports,

TABLE XXV
PATIENTS WITH WALDENSTRÖM'S MACROGLOBULINEMIA

Author	No. of Cases	Normal	Marker	% Cells with Marker	Remarks
Benirschke et al. (1962)	1	—	1	25%	
Bottura et al. (1961)	1	—	1	51%	
Broustet et al. (1966)	16				
Brown et al. (1967)	1		1	4%	
De Nava (1969)	9	9	—	—	4 healthy relatives show the marker
Elves and Israels (1963)	1		1	2%	
Ferguson and McKay (1963)	2	1	1	5%	
German et al. (1961)	1		1		
Heni and Siebner (1963)	1	—	1		
Houston et al. (1967)	7	2	5	2–30%	
Lustman et al. (1968)	2	1	1	12%	1 mother of a son with a D/D translocation.
Petit et al. (1968)	1	1	1	10%	
Pfeiffer et al. (1962)	2	2	1	8%	
Spengler et al. (1966)	1	1	1		small F marker, monozygotic healthy twin negative
Total	46	17	16		

Note: Investigations concern either blood or marrow chromosomes.

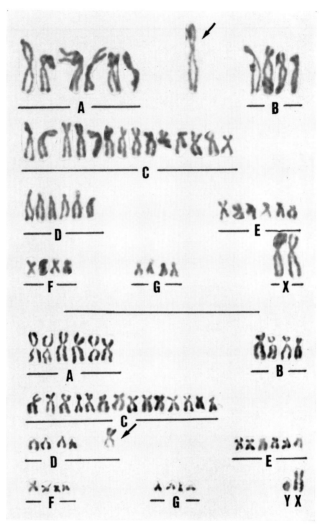

Figure 15. Upper karyotype is from female patient with Waldenström's macroglobulinemia. A supernumerary chromosome (arrow) was seen in 10 percent of lymphocyte mitoses. Karyotype below (son of the patient above) was carrier of D/D translocation and had normal phenotype. He had two normal daughters and his wife had no miscarriages. (Courtesy of Dr. F. Lustman et al., Acta Clin Belg, 23:67, 1968.)

illustrate the lack of specific significance of the so-called W-chromosome.

There are also other important theoretical discrepancies. Waldenström's macroglobulinemia may be transmitted as a dominant trait (McKusick, 1968): in such a case, the chromosome abnormality, or eventually the "tendency" to aneuploidy should also be transmitted. Brown et al. (1967) supported this assertion, showing a patient with macroglobulinemia and the marker, and four of his healthy relatives having at least one cell with the same marker. However, an opposite view is offered by Spengler et al. (1966). These authors could demonstrate the presence of an abnormal cell line in one monozygotic twin with macroglobulinemia, but not in the co-twin who was not affected. This strongly suggests that the abnormality is acquired and represents analogy to the cases of monozygotic twins, one with Ph^1 positive CML, the other healthy without Ph^1.

These opposite views shall be discussed in another chapter in greater detail together with the case of Lustman et al. (1968), in which a woman with Waldenström's macroglobulinemia and a larger marker in 12 percent of her blood cells had a healthy son in whom, however, an abnormal peak of gammaglobulins was demonstrated. He showed also, a patient with congenital chromosome abnormality, his karyotype being 45, XY, D-D-, t (Dq, Dq) (see Fig. 15).

MULTIPLE MYELOMA

According to Das and Aikat (1967) multiple myeloma is a progressive proliferative disorder of plasma cells and their precursors, and is generally regarded as a neoplastic process, fundamentally akin to leukemias and lymphomas. As in almost all neoplastic or suspected neoplastic processes, chromosomes have been investigated in this disease. Results have not been uniform and range from normal to the presence of markers or very marked aneuploidy (Table XXVI). In almost all cases, however, a normal cell line is found together with an abnormal one, and in Table XXVI the terms *hypo, pseudo* or *hyper* indicate the presence of an abnormal cell line, not meaning, however, that *all* cells are abnormal. More than one half of all cases show a marker chromosome. The marker, however, is different from report

TABLE XXVI
PATIENTS WITH MULTIPLE MYELOMA

Authors	No. of Cases	Hypo	Pseudo	Diplo	Hyper	Marker	Remarks
Baikie et al. (1959)	2	—	—	2	—	—	No marker
Bottura (1963)	5	1	1	3	—	1	pseudodiploid cell line shows marker
Castoldi et al. (1963)	1	1	—	—	—	—	no marker
Das and Aikat (1967)	5	2	—	1	2	—	no marker
De Nava (1969)	1	—	—	1	—	—	no marker
Houston et al. (1967)	17	—	10	7	—	10	pseudodiploid cell lines show a marker
Lewis et al. (1963)	3	—	—	—	3	3	marker in the hyperploid cell lines
Richmond et al. (1961)	3	—	—	3	—	—	no marker
Tassoni et al. (1967)	14	—	—	6	8	8	marker in hyperploid cells
Total	51	4	11	23	13	23	

to report: at times it is a large submetacentric chromosome (Houston et al., 1967) or a large acrocentric the size of a B group chromosome (Tassoni et al., 1967).

As in Waldenström's macroglobulinemia, a large number of mitoses must be counted in order to demonstrate the presence of an eventual marker, which may be seen either in the blood or bone marrow. The marker is not different from that seen in another gammopathy, macroglobulinemia. Thus Houston et al. (1967) proposed use of the term "MG chromosome" (for monoclonal gammopathy) instead of "W chromosome." However, this term is seldom used.

One patient in Houston's series exhibited CML and an abnormal gammaglobulin in the blood. Some cells had two markers: the Ph^1 and the "MG" chromosome. This could mean that the leukemic cells are capable of producing the abnormal globulin. Another described by Lewis et al. (1963) was an XO/XX mosaic. The marker chromosome was present in the XO cells, which therefore appeared pseudodiploid.

In summary, as in the case of Waldenstrom's macroglobulinemia, there is no unequivocal proof of a specific marker for multiple myeloma. Nevertheless, there is strong suggestion, theoretically very satisfactory, that such a marker might exist and could be related to the disease.

CONCLUSIONS

The following conclusions may be drawn:

1. Two forms of anemia may be considered: one is a preleukemic sign. In this form, malignancy develops rapidly and patients are classified as leukemic. The other may be called "true anemia": only these patients are herein considered.

2. In general, "true anemias" do not show chromosome abnormalities. Nowell's (1971) comments are germane:

> In non-irradiated patients with aneuploidy, the risk of developing clinical leukemia within the next few months is great, and patients who show such a progression probably had subclinical leukemia already present in the bone marrow at the time of study. If, however, frank leukemia does not appear within three months, preleukemic patients with marrow chromosome abnormalities are perhaps thereafter at no greater risk than comparable patients without such changes.

3. In Fanconi's anemia, a significant increase of rearrangements of nonspecific chromosomes can be observed. This could be linked with the high frequency of leukemia in this disease.

4. In Waldenström's macroglobulinemia as well as in multiple myeloma, a marker chromosome has been repeatedly observed in more than 50 percent of cases. This marker, however, is of changing morphology; although usually reported as large, being the size of an A or B chromosome, the position of the centromere may vary. When present, the marker has been observed in 1 percent to 51 percent of all mitoses (blood and bone marrow). The term W-chromosome or MG chromosome has been proposed. The cells carrying the chromosome abnormality are presumably those producing the abnormal protein.

5. Chromosome studies in clonal gammopathies certainly appear rewarding. When possible, both blood and bone marrow should be examined and at least 70 or 80 mitoses counted. Karyotypes must be carefully checked since the marker can appear in pseudodiploid cells.

6. Contradictory opinions do not allow us to conclude at present whether the W or MG chromosome is inherited or acquired. This specific point needs further investigation.

REFERENCES

Astaldi, G.; Strosselli, E.; Sauli, S.: Le cellule emiche nella ricerca citogenetica. *Haematologica, 47* (suppl):1, 1962.

Baikie, A. G.; Court Brown, W. M.; Jacobs, P. A.; Milne, J. S.: Chromosome studies in human leukemia. *Lancet, 2*:425, 1959.

Benirschke, K.; Brownhill, L.; Ebaugh, F. G.: Chromosomal abnormalities in Waldenström's macroglobulinemia. *Lancet, 1*:594, 1962.

Bottura, C.; Ferrari, I., Veiga, A. A.: Chromosome abnormalities in Waldenström's macroglobulinemia. *Lancet, 1*:1170, 1961.

Bottura, C.: Chromosome abnormalities in multiple myeloma. *Acta Haemat, 30*:274, 1963.

Broustet, A.; Hartmann, L.; Moulinier, J.; Staeffen, J.; Moretti, G.: Étude cytogénétique de 18 cas de maladie de Waldenström. *Nouv Rev Franc Hemat, 7*:809, 1967.

Brown, A. K.; Elves, M. W.; Gunson, W. H.; Pell-Ilderton, R.: Waldenström's macroglobulinemia. *Acta Haemat, 38*:184, 1967.

Castoldi, G. L.; Ricci, N.; Punturieri, E.; Bosi, L.: Chromosomal imbalance in plasmacytoma. *Lancet, 1*:829, 1963.

Chapelle, A. de la, and Grasbeck, R.: Normal mitotic activity and karyotype of leukocytes from pernicious anemia patients cultured in vitamin B_{12} deficient medium. Nature, 197:607, 1963.

Court Brown, W. M.; Jacobs, P. A.; Doll, R.: Interpretation of chromosome counts made on bone marrow cells. Lancet, 1:160, 1960.

Das, K. C., and Aikat, B. K.: Chromosomal abnormalities in multiple myeloma. Blood, 30:738, 1967.

De Nava, C.: Les anomalies chromosomiques au cours des hémopathies malignes et non malignes. Monogr Ann Genet, (Paris) l'Expansion, ed., 1969, vol. 1.

Elves, M. W. W., and Israels, M. C. G.: Chromosomes and serum proteins: a linked abnormality. Brit Med J, 2:1024, 1963.

Ferguson, J., and McKay, I. R.: Macroglobulinemia with chromosomal anomaly. Aust Ann Med, 12:197, 1963.

Ford, C. E.; Jacobs, P. A.; Lajtha, L. G.: Human somatic chromosomes. Nature, 181:1565, 1958.

Fortezza Bover, G., and Baguena Candela, R.: Analisis cytogenetica de un caso de anemia perniciosa antes y despues del tratamiento. Rev Clin Espan, 88:251, 1963.

German, J. L.; Bird, C. E.; Bearn, A. G.: Chromosomal abnormalities in Waldenström's macroglobulinemia. Lancet, 2:48, 1961.

Heath, C. W.: Cytogenetic observations in vitamin B_{12} and folate deficiency. Blood, 27:800, 1966.

Heni, F., and Siebner, H.: Chromosomenveränderung bei der Makroglobulinaemie Waldenström. Klin wschr, 40:342, 1962.

Houston, E. W.; Ritzman, S. E.; Levin, W. C.: Chromosomal aberrations common to three types of monoclonal gammopathies. Blood, 29:214, 1967.

Kiossoglou, K. A.; Mitus, W. J.; Dameshek, W.: Chromosomal aberrations in pernicious anemia. Study of three cases before and after therapy. Blood, 25:662, 1965.

Knospe, W., and Gregory, S. A.: Smoldering acute leukemia. Clinical and cytogenetic studies in six patients. Arch Intern Med, 127:910, 1971.

Leeksma, C. H. W.; Friden-Kill, L.; Brommer, E. J. P.; Neuberg, C. W.; Kerkhofs, H.: Chromosomes in premyeloid leukemia. Lancet, 2:1299, 1965.

Lewis, F. J. W.; Fraser, I. L.; MacTaggart, M.: An abnormal chromosomal pattern in myelomatosis. Lancet, 2:1013, 1963.

Lustman, F.; Stoffes-DeSaint Georges, A.; Ardichvili, D.; Koulischer, L.; Demol, H.: La macroglobulinémie de Waldenström. Acta Clin Belg, 23:67, 1968.

McKusick, V. A.: Mendelian Inheritance in Man, 2nd ed. Baltimore, The John Hopkins Press, 1968, vol. 1.

Petit, P.; Vryens, R.; Cauchie, C.; Koulischer, L.; Anomalie chromosomique du tissue ganglionnaire dans la macroglobulinémie de Waldenström, Acta Clin Belg, 23:182, 1968.

Pfeiffer, R. A.; Kosenow, W.; Baumer, A.: Chromosomenuntersuchungen an

Blutzellen eines Patienten mit Makroglobulinämie. Waldenström. *Klin Wschr, 40*:342, 1962.

Pierre, R. V., and Hoagland, H. C.: 45, X cell lines in adult men: loss of Y chromosome, a normal aging phenomenon? *Mayo Clin Proc, 46*:52, 1971.

Richmond, H. G.; Ohnuki, Y.; Pomerat, C. M.: Multiple myeloma—an in vitro study. *Br J Cancer, 15*:692, 1961.

Rowley, J. D.; Blaisdell, R. K.; Jacobson, L. O.: Chromosome studies in preleukemia. I. Aneuploidy of group C chromosomes in 3 patients. *Blood, 27*:782, 1966.

Spengler, G. A.; Siebner, H.; Riva, G.: Chromosomal abnormalities in macroglobulinemia Waldenström: discordant findings in uniovular twins. *Acta Med Scand 445* (suppl):132, 1966.

Tassoni, E. M.; Durant, J. R.; Becker, S.; Kravitz, B.: Cytogenetic studies in multiple myeloma: A study of fourteen cases. *Cancer Res, 27*:806, 1967.

Whang-Peng, J.; Henderson, E. S.; Knutsen, T.; Freireich, E. J.; Gart, J. J.: Cytogenetic studies in acute myelocytic leukemia with special emphasis on the occurrence of Ph[1] chromosome. *Blood, 36*:448, 1970.

Chapter XI

MALIGNANT DISEASES OF THE LYMPHORETICULAR SYSTEM

THE MALIGNANT DISEASES of the lymphoreticular system include follicular lymphoma, Burkitt's lymphoma, lymphosarcoma with or without leukemia, Hodgkin's disease, and reticulum cell sarcoma (Willis, 1967). Lymphoreticular neoplasias seem related to several genetic diseases, such as Chediak-Higashi syndrome, Wiskott-Aldrich syndrome, and in fact, many severe congenital immunologic deficiencies (Miller, 1966). Lack of correlation between histology and cytogenetics in these diseases has been pointed out by Millard (1968). The presence of a marker has been reported in five of nine cases of Burkitt's lymphoma by Jacobs et al. (1963); on the other hand, claim for a specific marker in malignant lymphomas was made by Spiers and Baikie (1966) who observed an 18p- chromosome in a few patients. However, further research failed to demonstrate the presence of a specific marker. Millard (1968) suggested that there was possible preferential involvement of chromosomes No. 18, but this has not been proven.

HODGKIN'S DISEASE

According to Smith et al. (1963), "Hodgkin's disease is a fatal disease in which there is a chronic, progressive, and painless enlargement of the lymph nodes and lymphoid structures in one or more regions of the body." Microscopic observation shows (Willis, 1967) a diffuse hyperplasia and proliferation of large reticuloendothelial cells —some of these assuming a giant form, both mononucleated and multi-nucleated—the so-called Reed-Sternberg cells. The most significant cytogenetic studies are those in which lymph node tissue has been studied (Table XXVII). Very often, special investigation of Reed-Sternberg cells has been made. Chromosome counts show two

TABLE XXVII
CHROMOSOME FINDINGS IN LYMPH NODES OF PATIENTS WITH HODGKIN'S DISEASE

Authors	No. of Cases	Hypo	Pseudo	Diplo	Hyper	Remarks
Baikie and Atkin (1965)	2	—	—	—	2	1 hypotriploid cell line
Galan et al. (1963)	1	—	—	—	1	2n = 83 (hypotetraploid)
Miles et al. (1966)	5	—	2	1	2	enhancement of secondary constriction in chr. 9
Ricci et al. (1962)	1	—	1	—	—	marker (iso E?)
Seif and Spriggs (1967)	6	—	—	2	2	2 near tetraploid, 2 "dubious" not included in the table (few mitoses)
Spiers and Baikie (1968)	5	—	2	3	—	18p—
Spriggs and Boddington (1962)	1	—	—	—	1	2n = 83 (hypotetraploid)

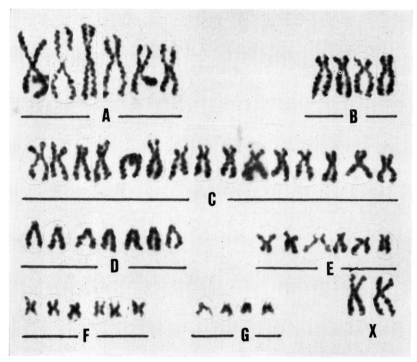

Figure 16. Chromosomes of a spleen puncture of a female patient with Hodgkin's disease, observed after 20 hours of incubation *in vitro*. There are 50 chromosomes, with one supernumerary C, one supernumerary D and two supernumerary F (50, XX, C+, D+, 2 F+). By this conventional staining technique, no structural rearrangements were observed. (Courtesy of Dr. Verhest, Department of Pathology, Institut Jules Bordet, Brussels.)

trends: diploidy or polyploidy. Polyploid mitoses may represent Reed-Sternberg cells as suggested by Spriggs and Boddington (1962), Ricci *et al.* (1962), and Galen *et al.* (1963). Studies on lymphocytes from the peripheral blood have shown poor response to phytohemagglutinin. Lawler *et al.* (1967) could count only 89 mitoses in 20 patients. Nineteen were aneuploid (12 being contributed by two patients) and nonspecific abnormalities were observed. Millard (1968), observing normal chromosomes in blood cultures of patients with malignant lymphoma, found abnormal karyotypes in cells from the lymph nodes. Deletion of the short arms of chromosome 18 has been reported by Spiers and Baikie (1966) in two cases of Hodgkin's disease. These authors even proposed the name of *Melbourne* or M^1 *chromosome* for this marker. However, it has not been observed in all cases (see Fig. 16) and certainly cannot be compared to the Ph^1 chromosome of chronic myeloid leukemia. The statement that an abnormality of chromosome 18 may be more often involved in this group of diseases (Seif and Spriggs, 1967; Millard and Seif, 1967) needs further proof before being accepted.

LYMPHOMAS AND LYMPHOSARCOMAS
(other than Burkitt's lymphoma)

As in other lymphoreticular neoplasias, the most significant results are those obtained by cytogenetic analysis of lymph nodes. Tjio *et al.* (1963) studied the chromosomes of their patient in blood and in lymph node aspirates: blood lymphocytes yielded normal chromosomes while lymph node cells showed a pseudodiploid cell line. As for Hodgkin's disease, chromosome counts fall mainly in the diploid range (Table XXVIII). The following remarks may be relevant for understanding of the results. In one example (Baikie and Atkin, 1965), a pseudodiploid stemline coexisted with a normal cell line. In the case of Sasaki *et al.* (1965), chromosome studies were done on peritoneal effusion and showed the presence of a marker. In the series of Spiers and Baikie (1966), five patients were considered to have lymphomas and six patients had follicular lymphoma. Finally, the patient of Kajii *et al.* (1968) belonged to a family in which three other brothers were affected by the disease ("familial lymphoma") but no other member of the family was studied.

Counts were mainly near diploid value. The hyperploid cell lines chiefly showed an increase by one or a few chromosomes. Only two of ten cases had polyploid cell lines, mainly in the range of tetraploidy. No specific abnormalities could be demonstrated. Millard and Seif (1967) observed structural rearrangements of chromosomes 18, mainly deletions of either the long or short arms in many mitoses. But they were not able to prove the presence of an abnormal cell line in any of their cases; thus their results are classified as normal in Table XXVIII.

BURKITT'S LYMPHOMA
(See chapter on viruses and chemical clastogens.)

TABLE XXVIII
CHROMOSOME FINDINGS IN LYMPHOMA AND LYMPHOSARCOMA

Author	No. of Cases	Hypo	Pseudo	Diplo	Hyper	Remarks
Baikie and Atkin (1965)	2	—	1	—	1	hyper: 2n = 49 or 50
Fitzgerald and Adams (1965)	6	—	—	5	1	rapid fatal evolution of patient with aneuploid cell line
Kajii et al. (1968.	1	—	1	—	—	pseudo: 3—, 18—, Y—, 16+, Bq—
Miles et al. (1966)	2	1	1	—	—	
Millard and Seif (1967)	6	—	—	6	—	sporadic cells with abnormal 18
Sandberg et al. (1964)	6	—	—	4	2	hyper: 2n = 47-51; 47-84
Sasaki et al. (1965)	2	—	1	—	1	pseudo: B marker (peritoneal effusion)
Spiers and Baikie (1968)	11	1	—	5	5	1 hyper polypl.
Tjio et al. (1963)	1	—	1	—	—	pseudo: C—, D—, A+, G+
Total	37	2	5	20	10	

Note: More than half of the cases exhibit normal karyotype.

TABLE XXIX
CHROMOSOME FINDINGS IN 6 CASES OF RETICULOSARCOMAS

Author	No. of Cases	Hypo	Pseudo	Diplo	Hyper	Polypl.	Remarks
Bauke and Schöffling (1968)	1	—	—	—	—	2	hypopentaploid
Spiers and Baikie (1968)	5	—	—	1	1	2	

RETICULOSARCOMAS

For Willis (1967), "reticulum cell sarcomas more strictly concern tumors in which the predominant or only line of differentiation of the neoplastic cells is towards reticular connective tissue, i.e. to fibrifying tumors allied to fibrosarcomas." The chromosomes in a few cases have been reported (Table XXIX). Polyploid cell lines have been observed in three cases. The patient of Bauke and Schöffling (1968) developed a secondary sideroachrestic anemia.

CHRONIC LYMPHOCYTIC LEUKEMIA

Chronic lymphocytic leukemia (CLL) is a malignant disease of the lymphoreticular system (Willis, 1967).

One of the first cytogenetic observations concerning CLL was that of Gunz et al. (1962). These authors described what they thought to be a specific chromosomal marker: a G-chromosome with deleted short arms. The marker was designated as Ch^1 or *Christchurch chromosome* since it has been discovered in Christchurch, New Zealand. However, further extensive research failed to confirm the specificity of the marker except for the case of Fitzgerald and Hamer (1969). It would thus appear that the Ch^1 chromosome is a familial characteristic, not directly associated with CLL.

With striking consistency, all reports on CLL mention normal diploid chromosomes. Berger and Parmentier (1971) collected 116 reported cases, all having normal karyotypes without any abnormal clone. These were patients of Baikie et al. (1960), Baker and Atkin (1965), Baserga et al. (1965 and 1966), Bayreuther (1960), Berger (1964), Colombiés et al. (1965), De Nava (1969, Court Brown (1964), Ducos and Colombiés (1968), Fitzgerald et al. (1966), Fitzgerald and Adams (1965), Fitzgerald and Gunz (1964), Ford (1960), Fraumeni et al. (1969), Gunz et al. (1962), Heni and Siebner (1964), Kinlough and Robson (1961), Koulischer (1968), Millard (1968), Nowell and Hungerford (1964), Oppenheim et al. (1965), Rozynkowa et al. (1968), Ruffié et al. (1966), Sandberg et al. (1960), Spiers and Baikie (1968).

Study of chromosomes in CLL raised some technical problems. Stimulation of leukemic lymphocytes by phytohemagglutinin (PHA) is impaired. After the usual incubation time (48 to 72 hours), few, if any, lymphocytes are stimulated (Schrek and Rabinowitz, 1963; Quaglino and Cowling, 1964; Winter et al., 1964). For Bernard et al. (1964), the transformation observed in "standard" cultures of lymphocytes of patients with CLL concerns the few normal lymphocytes still present. Moreover, the fact that leukemic lymphocytes are not stimulated by PHA could be considered as pathognomonic for CLL (Bernard et al., 1964). To overcome this difficulty, various modifications of the culture technique have been proposed, including a longer incubation period (6 rather than 3 days) and direct observation of lymph node cells as bone marrow aspirates. It is interesting to note that a minor degree of stimulation of lymphocytes by PHA has also been observed, among others, in Hodgkin's disease (Lawler et al., 1967).

Nonspecific anomalies of chromosomes, such as breaks, gaps, fragments, etc. have been observed in CLL, but the difference from those seen in cultures of healthy lymphocytes is not statistically significant (Berger and Parmentier, 1971). The finding by Fitzgerald (1965) that the small acrocentric chromosomes were shorter in patients with CLL than in normal individuals is possibly statistically significant in male patients (with 5 G group chromosomes including the Y), but not in female patients. Similar observations of shortened G group chromosomes have been reported in four patients with Marfan's syndrome (Källén and Levan, 1962). It could be assumed that the apparent cytogenetic "normality" in CLL is related to the nature of the disorder itself, which could be merely an "accumulative disease of immunologically incompetent lymphocytes" than a malignancy characterized by a rapid proliferation of abnormal cells (Dameshek, 1967).

CONCLUSION

1. Almost 60 percent of all cases of lymphoreticular neoplasms show diploid or pseudodiploid karyotypes. Polyploidy is not infrequent (10% of all cases) and may be linked with the presence of giant mono- or multinucleated cells in the tumors (as in the case of Reed-Sternberg cells in Hodgkin's disease).

2. No specific marker has been demonstrated in any disease of this group. Claims for a specific marker in Hodgkin's disease, the Melbourne (M^1) chromosome and Christchurch (Ch^1) chromosome in

chronic lymphocytic leukemia are premature, as is the statement that chromosome 18 is preferentially involved in neoplasia of the lymphoreticular system.

3. Chromosome abnormalities observed in Burkitt's lymphoma are different from experimental, virus-induced abnormalities.

REFERENCES

Baikie, A. G.; Court Brown, W. M.; Jacobs, P. A.: Chromosome studies in leukemia. *Lancet, 1*:168, 1960.

Baker, M. C., and Atkin, N. B.: Chromosomes in short-term cultures of lymphoid tissue from patients with reticulosis. *Br Med J, 1*:770, 1965.

Baserga, A.; Castoldi, G. L.; Franceschini, F.: Étude chromosomomique de la leucémie lymphocytair chronique. *Schweiz Med Wochn, 96*:1220, 1966.

Bauke, J., and Schöffling, K.: Polyploidy in human malignancy. *Cancer, 22*:686, 1968.

Bayreuther, K.: Chromosomes in primary neoplastic growth. *Nature, 186*:6, 1960.

Berger, R.: Contribution a l'étude cytógénetique des leucémies humaines. Thèse, Faculté de Médecine, Paris, 1964.

Berger, R.: Contribution à l'étude cytógénetique des leucémies humaines. somes. *Nouv Rev Franc Hémat, 11*:261, 1971.

Bernard, C.; Geraldes, A.; Boiron, M.: Action de la phytohémagglutinine in vitro sur les lymphocytes de leucémies lymphoides chroniques. *Nouv Rev Franc Hémat, 4*:69, 1964.

Colombiès, P.; Ducos, J.; Ruffié, J.; Salles-Mourlan, A. M.: Existe-t-il des anomalies chromosomiques au cours des leucémies lymphocytaires chroniques? Étude de 16 cas. *Rev Franc Etud Clin Biol, 10*:525, 1965.

Court Brown, W. M.: Chromosomal abnormality and chronic lymphatic leukaemia. *Lancet, 1*:986, 1964.

Dameshek, W.: Chronic lymphocytic leukemia—an accumulative disease of immunologically incompetent lymphocytes. *Blood, 29*:566, 1967.

De Nava, C.: Les anomalies chromosomiques au cours des hémopathies malignes et non malignes. *Monogr Ann Génét* (Pairs), l'Expansion ed., 1969, vol. 1.

Ducos, J., and Colombiès, P.: Chromosomes in chronic lymphocytic leukemia. *Lancet, 1*:1038, 1968.

Fitzgerald, P. H.: Abnormal length of the small acrocentric chromosomes in chronic lymphocytic leukemia. *Cancer Res, 25*:1904, 1965.

Fitzgerald, P. H., and Gunz, F. W.: Chromosomal abnormality and chronic lymphocytic leukemia. *Lancet, 2*:150, 1964.

Fitzgerald, P. H., and Adams, A.: Chromosome studies in chronic lymphocytic leukemia and lymphosarcoma. *J Nat Cancer Inst, 34*:827, 1965.

Fitzgerald, P. H.; Crossen, P. E.; Adams. A. C.; Sharman, C. V.; Gunz, F. W.: Chromosome studies in familial leukemia. *J Med Genet, 3*:96, 1966.

Fitzgerald, P. H., and Hamer, J. W.: Third case of chronic lymphocytic leukemia in a carrier of the inherited Ch^1 chromosome. *Br Med J, 3*:752, 1969.

Ford, C. E.: The chromosomes of normal human somatic and leukemic cells. *Proc Roy Soc Med, 53*:491, 1960.

Fraumeni, J. F.; Vogel, C. L.; Devita, V. T.: Familial chronic lymphocytic leukemia. *Ann Intern Med, 71*:279, 1969.

Galan, H. M.; Lida, E. J.; Kleisner, E. H.: Chromosomes of Sternberg-Reed cells. *Lancet, 1*:335, 1963.

Gunz, F. W.; Fitzgerald, P. H.; Adams, A.: An abnormal chromosome in chronic lymphocytic leukemia. *Br Med J, 2*:1097, 1962.

Heni, F.; Siebner, H.: Chromosomal abnormality and chronic lymphocytic leukemia. *Lancet, 1*:1109, 1964.

Jacobs, P. A.; Tough, I. M.; Wright, D. H.: Cytogenetic studies in Burkitt's lymphoma. *Lancet, 2*:1144, 1963.

Kajii, T.; Neu, R. L.; Gardner, L. I.: Chromosome abnormalities in lymph node cells from patient with familial lymphoma. *Cancer, 22*:218, 1968.

Källén, B., and Levan, A.: Abnormal length of chromosomes 21 and 22 in four patients with Marfan's syndrome. *Cytogenetics, 1*:5; 1962.

Kinlough, M. A., and Robson, H. N.: Study of chromosomes in human leukemia by a direct method. *Br Med J, 2*:1052, 1961.

Koulischer, L.: Contribution a l'etude des chromosomes dans les leucémies humaines. Thèse, Brussels, 1968.

Lawler, S. D.; Pentycross, C. R.; Reeves, B. R.: Lymphocyte transformation and chromosome studies in Hodgkin's disease. *Br Med J, 3*:704, 1967.

Miles, C. P.; Geller, W.; O'Neill, F.: Chromosomes in Hodgkin's disease and other malignant lymphomas. *Cancer, 19*:1103, 1966.

Millard, R. E.: Chromosome abnormalities in the malignant lymphomas. *Europ J Cancer, 4*:97, 1968.

Millard, R. E., and Seif, G. Chromosomes in malignant lymphomas. *Lancet, 1*:781, 1967.

Miller, R. W.: Relation between cancer and congenital defects in man. *N Engl J Med, 275*:87, 1966.

Nowell, P. C., and Hungerford, D. A.: Chromosome changes in human leukemia and a tentative assessment of their significance. *Ann NY Acad Sci, 113*:654, 1964.

Oppenheim, J. J.; Whang, J.; Frei, E.: Immunologic and cytogenetic studies of chronic lymphocytic leukemia cells. *Blood, 26*:121, 1965.

Quaglino, D., and Cowling, D. C.: Cytochemical studies on cells from chronic lymphocytic leukaemia and lymphosarcoma cultured with phytohaemagglutinin. *Br J Haemat, 10*:358, 1964.

Ricci, N.; Punturieri, E.; Bosi, L.; Castoldi, G. L.: Chromosomes of Sternberg-Reed cells. *Lancet, 2*:564, 1962.

Rozynkowa, D.; Marczak; Rupniewska, Z.: A chromosome abnormality in lymphatic leukemia. *Humangenetik, 6*:300, 1968.

Ruffié, J.; Ducos, J.; Bierme, R.; Colombies, P.; Salles-Mouran, A. M.: Chromosomes in chronic lymphocytic leukemia. *Lancet, 2*:227, 1966.

Sandberg, A. A.; Koepf, G. F.; Crosswhite, L. H.; Hauschka, T. S.: The chromosome constitution of human marrow in various developmental and blood disorders. *Am J Hum Genet, 12*:231, 1960.

Sandberg, A. A.; Ishihara, J.; Kikuchi, Y.; Crosswhite, L. M.: Chromosomal differences among the acute leukemias. *Ann NY Acad Sci, 113*:663, 1964.

Sasaki, M. S.; Sofuni, T.; Makino, S.: Cytological studies in tumors. XLII. Chromosome abnormalities in malignant lymphomas of man. *Cancer, 18*:1007, 1965.

Schrek, R., and Rabinowitz, Y.: Effects of phytohaemagglutinin on rat and normal and leukemic human blood cells. *Proc Soc Exptl Biol Med* (NY), *113*:191, 1963.

Seif, G. S. F., and Spriggs, A. I.: Chromosomes changes in Hodgkin's disease. *J Nat Cancer Inst, 39*:557, 1967.

Smith, N. J.; Vaughan, V. C.; Diamond, C. K.: The lymphatic system. In Nelson, W. E.: *Textbook of Pediatrics.* Philadelphia, London, W. B. Saunders, 1963, vol. 1.

Spiers, A. S. D., and Baikie, A. G.: Cytogenetic studies in the malignant lymphomas and related neoplasms. *Cancer, 22*:193, 1968.

Spriggs, A. I., and Boddington, M. M.: Chromosomes of Sternberg-Reed cells. *Lancet, 2*:153, 1962.

Tjio, H. J.; Marsh, J. C.; Whang, J.; Frei, E.: Abnormal karyotype findings in bone marrow and lymph node aspirates of a patient with malignant lymphoma. *Blood, 22*:178, 1963.

Winter, G. C. B.; Osmond, D. G.; Yoffey, J. M.; Mahy, D. J.: Leukocyte cultures with phytohemagglutinin in chronic lymphatic leukemia. *Lancet, 2*:563, 1964.

Willis, R. A.: *Pathology of Tumors,* 4th ed. London, Butterworths Medical Publications, 1967.

PART III
SOLID TUMORS

Chapter XII

FEMALE REPRODUCTIVE SYSTEM

IN 1970, almost one-half (42.5%) of estimated new cases of cancer in women of all ages were tumors of the breast and genitals (Silverberg and Grant, 1970). Breast cancer was most frequent, followed by cancer of the cervix uteri, the latter having a steady decline attributed to regular checkup with smear tests. Five-year survival rates, adjusted for normal life expectancy were 61 percent for breast cancer, 65 percent for uterine cancer, and 30 percent for ovarian cancer (Silverberg and Grant, 1970).

Because of the large number of case reports, especially of carcinoma of the cervix, we have attempted statistical analyses of modal values which we believe will be more meaningful here than in the case of tumors discussed in previous chapters.

OVARIAN TUMORS

Ovarian cancers account for about 20 percent of all tumors of the female reproductive organs. They are generally classified as neoplasms derived from (a) celomic germinal epithelium, (b) connective tissue, (c) specialized differentiation of cortical stroma, (d) urogenital epithelium, and (e) all three germ layers (teratomas).

All these tumors have been analyzed chromosomally, most frequently cystadenocarcinomas.

Dermoid cyst of the ovary exhibited a normal karyotype in 24 analyzed metaphases (Toews, 1968). In a "vegetant cyst" of the ovary, abnormal metaphases were found by de Grouchy *et al.* (1963), cells being of two distinct modes: 45 and 88 to 90 with counts ranging 36 to 94 in 57 cells.

In truly malignant lesions the picture was different. Distinct markers were observed by Ruffié *et al.* (1964), Hansen-Melander *et al.* (1956) and Atkin and Baker (1966). These marker chromosomes were seen in ovarian cystadenoma and adenocarcinoma and occurred also in their peritoneal effusions. Despite some indication of consistency in findings (Atkin and Baker, 1966), the hypothesis of specificity of a long marker could not be substantiated with significant data. Fraccaro's finding (1965) of 57 percent breaks in cystadenoma was not repeatedly observed by other authors.

Reviewing the literature, we have analyzed 79 case reports from effusions and 54 chromosomal analyses of solid ovarian malignant tumors. Included were only those which contained modal cells. Comparing the modal values of tumor effusions and solid tumor cells, no significant difference was noted. Mean modal number in effusions was 55.2 ($S^2 = 250.1$) and in solid tumors 56.4 ($S^2 = 217.0$), the mean difference being 1.18 with a standard deviation of 2.7 and a t score of 0.434.

We have observed that the mean number of cells with a modal chromosome number was higher in effusions (35.8%) than in solid tumors (26.2%) but the difference was not outstanding (standard deviation of difference of means $= 1.323$, d.f. $= 51.1$).

The above statistical data included all types of ovarian malignant tumors with the exception of teratomas.

Review of 15 reported cases of *ovarian teratomas* showed almost consistently a normal diploid mode, rarely with deviation in morphology of chromosomes. The mean modal number was 46.133 ($S^2 = 0.124$).

CARCINOMA OF THE BREAST

Malignant tumors of the breast may arise from stratified squamous epithelium, glandular structures, or mesenchymal connective tissue, but the majority—over 90 percent—arise in ductal epithelium. It is generally agreed that in some families there is an inherited predisposition. Hormonal influence, especially estrogens, seems to play an etiological role. The influence of other possible factors such as viruses was neither documented nor disproved and lately has attracted much attention. In the United States, breast cancer is the most common tumor in females (Silverberg and Grant, 1970) representing about 18 percent of all female cancer.

Chromosomes of breast cancer have been analyzed frequently. More than 100 case reports were characterized with modes. Large markers

(Castoldi, 1968; Ruffié, 1964), and rings (Katayama and Masukawa, 1968) were observed in some but without any consistent repetition. For instance, Toews et al. (1968) examined four cases of *benign cystic mammary disease* finding all metaphases to have a normal set of 46 chromosomes. The same findings were made in two cases of *in situ lobular* carcinoma. All specimens were examined after two hours incubation with colchicine. Emson and Kirk (1967) measured the DNA content of interphase cells by the two-wave-lengths microspectrophotometric method in *lactational hyperplasia* and *cystic hyperplasia*. Only normal DNA values were found thus failing to shed additional light on the old controversy concerning the question of whether cystic hyperplasia of the breast is premalignant.

We reviewed 104 reports of chromosomal analyses of breast cancer containing modal cells, 54 from effusions and 50 from solid tumors. The mean modal number was higher in effusions (62.3, $S^2 = 518.3$) than in solid tumors (53.3, $S^2 = 212.3$). The difference of means was 9.088 and t score $= 2.403$, d.f. $= 102$. Hyperdiploidy thus prevailed.

CANCER OF THE CERVIX UTERI

The second most frequent malignancy of women, carcinoma of the uterine cervix, is also the tumor exhibiting the most rapid decline as a cause of death. This is due to its comparatively accessible location and improved methodology of early detection. For these reasons large amounts of cytogenetic data have been collected at different stages in this type of malignancy. Especially valuable have been findings in premalignant stages such as dysplasias, atypias, and carcinoma in situ.

Twenty-nine cases of lesions classified as *dysplasias* and *atypias* were collected from the literature with the following findings: Auersperg et al. (1967) examined three cases of minimal and moderate dysplasia with cells of the normal diploid chromosomal set. One "marked" dysplasia showed a wide range in counts and a mode of 70 to 80 chromosomes. Richart and Wilbanks (1966) and Richart and Corfman (1964) described 11 cases of dysplasia classified as mild to severe, all of which had a modal number of 46. However, in none of the specimens did all metaphases have 46 chromosomes: the proportion of aberrant metaphases ranged from 10 to 46 percent and did not correspond to the severity of dysplasia in the sense that the more severe the histological aberration, the higher the percentage of nonmodal metaphases found. Also, the range of counts did not get wider with increasing severity of the lesion.

Among six cases of dysplasia reported by Wakonig-Vaartaja (1965), four had a mode of 46 in only 34.6 to 52 percent of cells. Two other cases showed even more variability, displaying no definite modal number and having count ranges of 42 to 61 and 44 to 51. The trend to more disorganized karyotypes characteristic for malignant changes was also observed in dysplasia by other authors. Jones *et al.* (1967) found only one of 12 cases of cervical atypia with modal number 46 (85.2%). Boddington *et al.* (1965) observed three cases, all having abnormal chromosome numbers. In most reports, chromosomal counts did not reach high values and remained clustered around the diploid number with a narrow range.

Kirkland (1966) estimated that there is less than 50 percent agreement on histologic diagnosis of dysplasia and about 60 percent of carcinoma in situ of the cervix. Distinction between chromosome findings in carcinoma in situ and dysplasia is therefore likely to project this diagnostic uncertainty.

Thus, we see that in *carcinoma in situ* of the cervix some specimens show only slight deviation from normal diploidy. Generally, however, the chromosome numbers exhibit wider scatter. The many tetraploid and hypertetraploid karyotypes represent more advanced changes in comparison with findings in dysplasia and atypia. Less than one-half of the cases analyzed displayed any modal number. Another characteristic factor appears: histologically abnormal mitoses occur frequently while these were not observed in mere dysplasias.

In our analyses of 29 case reports of dysplasia and atypia compared with 31 reported cases of carcinoma in situ, the mean modal number for dysplasia and atypia was 54.9 ($S^2 = 325.9$) and for carcinoma in situ, 55.1 ($S^2 = 290.0$). The difference was not significant (t score $= 0.036$, St $(\bar{x}_1 - \bar{x}_2) = 4.52$). Graphic demonstration of the proportion of modal cells, range of counts, and difference in diploid and pseudodiploid modes were presented by Kirkland and Stanley (Figs. 17, 18, 19). Additional modes in carcinoma in situ characterized the difference from dysplasias and atypias.

Female Reproductive System 135

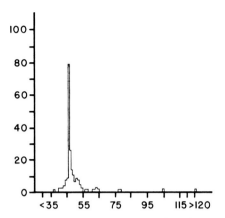

Figure 17. The distribution of chromosome numbers in cervical dysplasia exhibits a limited range of chromosome numbers. 186 cells of nine cases were analyzed showing mostly diploid chromosomal constitution. (Courtesy of Drs. Kirkland and Stanley, *Aust N Z J Obstet Gynaec*, 7:189, 1967.)

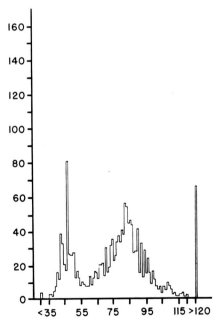

Figure 18. The distribution of chromosome numbers in 1681 cells of 67 cases of carcinoma in situ of the cervix. Note the wide range of aneuploidy and the second population of cells with chromosome numbers in the range 75 to 95. (Courtesy of Drs. J. A. Kirkland and M. A. Stanley, *Aust N Z J Obstet Gynaec*, 7:189, 1967.)

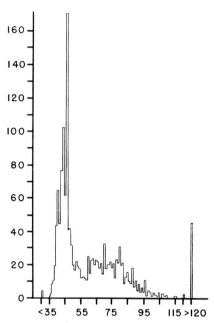

Figure 19. Chromosome distribution in invasive carcinoma in 1583 cells of 76 cases of the cervix uteri. There was a wide range in chromosome numbers and no second peak (see Fig. 18) was observed. (Courtesy of Drs. J. A. Kirkland and M. A. Stanley, *Aust N Z J Obstet Gynaec,* 7:189, 1967.)

In *invasive carcinoma,* a karyotypic difference between small cell types and large cell types was observed (Kirkland and Stanley, 1971; Atkin, 1964). The presence of a mode was more prominent in the small cell type than in the large cell type. It was further suggested that the small cell lesions are radioresistant. In small cell types, pseudodiploid karyotypes are frequently noted, a finding associated by several authors with onset of invasion. Not a vast disorganization of karyotype, but the emergence of pseudo- or near-diploid modal cell populations is characteristic of invasive forms of carcinoma of the cervix. Atkin and Richards (1962) and, later, Kirkland (1966) concluded that lesions with chromosomes close to the diploid mode or multiples of the haploid mode are more dangerous. Statistical significance was high and the findings were also supported by DNA content analysis by Atkin (1966) who, after an observation of 998 cases of carcinoma of the cervix, concluded that higher ploidy tumors had better prognosis compared with near diploid tumors.

In an attempt to express the difference in modal number between 60 reported cases of carcinoma of the cervix and 31 reported cases of carcinoma in situ of the cervix, we have compared their mean modal numbers. That of the former was 61.5 ($S^2 = 290.04$). The difference was significant (t score $= 1.2$, d.f. 89) and supports, in part, the hypothesis that there is a karyotypic difference associated with development of cervical cancer and its prognosis as follows:

In *dysplasia or atypia* a large proportion of cases are found to have cells with normal diploid karyotype, but also significant number of abnormal mitoses deviating from the diploid number. Pseudodiploid figures are found.

In *carcinoma in situ* a widely scattered number of chromosomes is found, many karyotypes being in the hypo- or hypertetraploid range. A secondary mode commonly appears.

In *invasive carcinoma* widely scattered chromosomal numbers occur, but in contrast to carcinoma in situ, secondary modes in the tetraploid region are rare and the main mode most frequently is around the diploid value. The tumors with widely scattered chromosomal numbers have better prognosis than those with near-diploid or pseudodiploid modes.

CARCINOMA OF THE CORPUS UTERI

Much less frequent a primary site than cancer of the cervix, cytogenetic findings in carcinoma of the corpus uteri markedly differ in several reports. Less often detected in the initial stages, cytogenetic analyses do not yield much information about changes of karyotype during the course of development of malignancy.

For invasive carcinoma of the corpus uteri the data are generally inconclusive, but show the typical picture of disorganized karyotype with modes as low as 28 (Richards and Atkin, 1960) and as high as tetraploidy (Makino *et al.*, 1964). Various markers were observed but there was no uniformity in the findings of different authors.

Analyzing 39 reported cases which were characterized by modal numbers and comparing the mean modal number for carcinoma of the corpus uteri (50.56) with that of carcinoma of the cervix (61.13), we found the modal value for the former to be significantly lower and nearer to diploidy ($\bar{x}_1 - \bar{x}_2 = 10.57$, t score $= 2.471$, d.f. $= 97$). Furthermore, the difference between modes in different specimens of cancer of the corpus uteri was smaller ($S^2 = 127.78$) than in carcinoma of the cervix ($S^2 = 628.52$) which have shown marked scatter.

One report of benign *fibromyoma uteri* was encountered (Fiocchi, 1967). A normal set of 46 chromosomes was found by the direct method.

CARCINOMA OF THE ENDOMETRIUM

Carcinoma of the endometrium develops in the vast majority of cases after menopause. It is five to seven times less frequent than carcinoma of the cervix. Prognosis is usually relatively favorable because bleeding frequently attracts the attention of the patient before there is extensive invasion of the tumor.

Of 67 recorded cases of carcinoma of the endometrium, 28 exhibited a clear mode. The mean modal number of cases with an apparent mode was 46.64 ($S^2 = 20.9$), thus not deviating markedly from the normal diploid set. However, about 60 percent of reported cases did not show any definite mode and chromosome counts were usually scattered around diploid and triploid values. Analyses from solid tumors and effusions did not display significant differences.

Normal metaphases with 46 chromosomes are usual in endometrial carcinoma. Their proportion decreases, however, with increasing invasiveness of the tumor.

Two cases of *atypical endometrial hyperplasia* were examined by Stanley and Kirkland (1968). One exhibited a normal karyotype and the other had a majority of pseudodiploid cells and six cells in the range of 38 to 46.

In this regard, it is worthwhile to review some of the published data on normal endometrial chromosome findings. During the last decade, it was repeatedly found that normal endometrium contains a considerable proportion of polyploid cells. Since the early reports, different authors have shown great variability in findings (Sachs, 1953; Walker and Boothroyd, 1954; Tjio and Puck, 1958; Takemura, 1960; Hughes and Czermely, 1965, 1966; Sherman, 1969; and others). Recently, the investigation of Rask-Madsen and Philip (1970) on 2,954 endometrial cell mitoses studied by the *direct method* have shed new light on the question of physiologic polyploidy. These authors found only 5.5 percent of aneuploid cells in their large sample, suggesting that the previously reported high percentage of polyploid cells were most likely due to technically unsuitable procedures.

CHORIONIC VILLI, HYDATIDIFORM MOLE, CHORIONEPITHELIOMA

Trophoblastic tumors are estimated to occur in Caucasians much less frequently than in Japanese and Chinese. Green (1964) estimated the frequency of hydatidiform mole to be about 1:2000 to 1:2500 pregnancies in Caucasians of Australia, while the estimate in a Taiwanese population was about 1:125. In South East Asia and India the frequency is about seven to ten times higher than in Europe or in the United States (Llewellyn and Jones, 1965). Moles are more frequent in mothers younger than 20 years and older than 39 years.

There is a suggested developmental continuity between abnormal chorionic villi and malignant chorionepithelioma in the following sequence: normal villi→transitional hydatidiform mole-like villi→hydatidiform mole→destructive hydatidiform mole→chorionepithelioma.

This continuity, at least between transitional villi and mole, seems to be supported by chromosomal studies by several authors. First, Sasaki *et al.* (1962) examined 2000 cells of *normal chorionic villi* by the direct method, finding 0.1 to 0.45 percent mitotic figures. Of 2000 cells examined in hydatidiform moles in the fourth month of pregnancy the proportion of mitotic figures was higher (0.9% to 1.7%). Further, a total of 552 mitoses from three cases (abortions) of apparently normal villi were also analyzed. In all three cases the mode was clearly 46 and polyploid cells were found in 0.9 percent, 1.0 percent, and 1.0 percent, respectively. Kawasaki (1968) also confirmed normal chromosomal findings.

The first apparent chromosome changes were detected in *abnormal transitional villi* by Makino *et al.* (1964). From three women in the first trimester of pregnancy, specimens of villi characterized by proliferation of trophoblasts and edematization of stromal cells with absence or extreme scantiness of blood vessels were obtained. In all three cases the chromosomal count ranged from 53 to 71 with a distinct mode at 69. All triploid cells showed three times the haploid autosomal number plus an XXY set of sex chromosomes. A total of 75 mitoses was analyzed.

Hydatidiform mole, in the early stages and even more in its destructive development, showed abnormal karotypes. However, with regard to modal number, this value generally did not greatly deviate from near diploidy. For example, of 17 cases reported to have modal cells, the mean mode was close to diploidy. Nine had a modal number of 46, one being pseudodiploid (Kawasaki, 1968; Serr, *et al.* 1968; Stolte, *et al.* 1960). In all nine cases, however, the proportion of aneuploid cells was higher than 16 percent except for one case with 1.4 percent aneuploid metaphases.

We found only four reported chromosomal analyses of chorionepithelioma (Kawasaki *et al.*, 1968; Serr *et al.*, 1969). All had near diploid abnormal stemlines except for one having 18 cells with normal 46 chromosomes and 4 aneuploid cells.

MISCELLANEOUS GYNECOLOGIC TUMORS

Adenocarcinoma of the vagina examined by Wakonig-Vaartaja and Hughes (1967) yielded 54 metaphases with abnormal chromosomes and a mode of 68. (*Squamous cell carcinoma of the vagina* was analyzed in three cases (Wakonig-Vaartaja, and Hughes 1967; Goodlin, 1962). One case was characterized with a modal number of 46. The remaining two showed abnormal chromosomes without a definite mode. *Adenocarcinoma of the Fallopian tube* (Goodlin, 1962) had an abnormal mode of 82 to 84. *Carcinoma in situ of the vulva* (Wakonig-Vaartaja, and Hughes 1967) had abnormal chromosomes but no modal cells.

A solid specimen of *Bowen's disease* was studied by Makino *et al.* (1959). A modal number of 56 to 59 was found in 36 percent of cells analyzed.

Of nonmalignant tumors, a solid specimen of *fibromyoma uteri* yielded only normal metaphases (Fiocchi, 1967).

SEX CHROMATIN IN BREAST CANCER

Another aspect related to abnormalities of chromosomes is the incidence of sex chromatin in interphase cancer cells. Several authors concluded that there are "masculine" and "feminine" types of breast cancer in females depending on the presence of Barr bodies (Ehlers and Hienz, 1958; Kimel, 1957). Furthermore, it was suggested that those tumors with a high frequency of sex chromatin have unfavorable prognosis and hormonal therapy does not meet with success (Shirley, 1967; Klug, 1969). This hypothesis was not confirmed by others (Scholl *et al.*, 1968). The evidence appears to be inconclusive, especially in the light of reasoning by Kallenberger *et al.* (1967) who explains the absence of sex chromatin as cellular dedifferentiation and consequently loss of diploidy—resulting in a severely malignant character of the cell.

REFERENCES

Atkin, N. B., and Richards, B. M.: DNA in human tumors as measured by microspectrophotometry of Feulgen stain: a comparison of tumours arising at different sites. *Br J Cancer, 10*:769, 1956.

Atkin, N. B.; Richards, B. M.; Ross, A. J.: The deoxyribonucleic acid content of carcinoma of the uterus: an assessment of its possible significance in relation to histopathology and clinical course, based on data from 165 cases. *Br J Cancer, 13*:773, 1959.

Atkin, N. B., and Klinger, H. P.: The superfemale mole. *Lancet, 2*:727, 1962.

Atkin, N. B., and Richards, B. M.: Clinical significance of ploidy in carcinoma of cervix. Its relation to prognosis. *Br Med J, 2*:1445, 1962.

Atkin, N. B., and Baker, M. C.: A nuclear protrusion in a human tumor associated with an abnormal chromosome. *Acta Cytol, 8*:431, 1964.

Atkin, N. B.: The influence of nuclear size and chromosome complement on prognosis of carcinoma of the cervix. *Proc Roy Soc Med, 59*:979, 1966.

Atkin, N. B., and Baker, M. C.: Chromosome abnormalities as primary events in human malignant disease. Evidence from marker chromosomes. *J Nat Cancer Inst, 36*:539, 1966.

Atkin, N. B.; Mattison, G.; and Baker, M. C.: A comparison of the DNA content and chromosome number of fifty human tumours. *Br J Cancer, 20*:87, 1966.

Atkin, N. B.; Baker, M. C.; and Wilson, S.: Stemline karyotypes of four carcinomas of the cervix uteri. *Am J Obstet Gynecol, 99*:506, 1967.

Amarose, A. P., and Baxter, D. H.: Chromosomal changes following surgery and radiotherapy in patients with pelvic cancer. *Obstet Gynecol, 25*:828, 1965.

Auersperg, N., and Hawryluk, A. P.: Chromosome observations on three epithelial-cell cultures derived from carcinoma of the human cervix. *J Nat Cancer Inst, 28*:605, 1962.

Auersperg, N.; Corey, M. J.; and Worth, A.: Chromosomes in preinvasive lesions of the human uterine cervix. *Cancer Res, 27*:1394, 1967.

Baker, M.: A chromosome study of seven near diploid carcinomas of corpus uteri. *Br J Cancer, 22*:683, 1968.

Bamford, S. B.: Mitchell, G. W. Jr.; David, J.; Sperber, A.; and Cassin, C.: Size variation of the late replicating X-chromosome in the leukocytes of individuals with hyperplastic and malignant lesions of uterine epithelium. *Acta Cytol, 13*:238, 1969.

Beischer, N. A.; Fortune, D. W.; and Fitzgerald, M. G.: Hydatidiform mole and coexistent foetus both with triploid chromosome constitution. *Br Med J, 3*:476, 1967.

Biswas, S., and Chowdhury, J. R.: Evaluation of chromosome morphology in human normal and cancer cervix cells in vitro. *Indian J Med Res, 56*:1595, 1968.

Boddington, M. M.; Spriggs, A. I.; Wolfendale, M. R.: Cytogenetic abnormalities in carcinoma in situ and dysplasias of the uterine cervix. *Br Med J, 1*:154, 1965.

Boyes, D. A.: Fidler, H. K.; Lock, D. R.: Significance of in situ carcinoma of the uterine cervix. *Br Med J, 1*:293, 1962.

Castoldi, G. L.; Scapoli, G. L.; and Spanedda, R.: Sull evoluzione clonale del cariotipo delle cellule neoplastiche in versamento pleurico da adenocarcinoma mammario. *Arch Ital Pat, 11*:3, 1968.

Corfman, P. A., and Richart, R. M.: Chromosome number and morphology of benign ovarian cystic teratomas. *N Engl J Med, 271*:1241, 1964.

Cox, L. W.; Stanley, M. A.; Harvey, N. D. M.: Cytogenetic assessment of radiosensitivity of carcinoma of the uterine cervix. *Obstet Gynecol, 33*:82, 1969.

Edmonds, H. W.: Genesis of hydatidiform mole: old and new concepts. *Ann NY Acad Sci, 30*:86, 1959.

Ehlers, P. N., and Hienz, H. A.: Zellkernmorphologisches Geschlecht und hormonelle Beeinflussbarkeit des Mammakarzinoms. *Langenbecks Arch, 288*:485, 1958.

Eick, J.; Emminger, A.; Strauss, C.; Mohr, U.; Wrba, H.: Cytogenetisch-karyologische Studien an klinischen behandelten gynäkologischen Tumoren. *Z Krebsforsch, 67*:205, 1965.

Emson, H. E., and Kirk, H.: Value of desoxyribonucleic acid (DNA) in evolution of carcinomas of the human breast. *Cancer, 20*:1248, 1967.

Fiocchi, E.: Studies of the chromosome patterns of some types of tumors of the female genital system. *Folia Hered Path, 16*:157, 1967.

Fischer, P.; Golob, E.; Holzner, J. H.: Chromosomenanzahl und DNS-Wert bei malignen Tumoren des weiblichen Genitaltraktes. *Z Krebsforsch, 68*:200, 1966.

Fischer, P.; Golob, E.; Holzner, J. H.: Zytogenetische Untersuchungen am Portioepithel bei positivem und bei zweifelhaltem Abstrichbefund. *Krebsarzt,* 22:289, 1967.

Fraccaro, M.; Mannini, A.; Tiepolo, L.; Zara, C.: High frequency of spontaneous recurrent chromosome breakage in an untreated human tumour. *Mutat Res,* 2:559, 1965.

Fraccaro, M.; Tiepolo, L.; Gerli, M.; Zara, C.: Analysis of karyotype changes in ovarian malignancies. *Panminerva Med,* 8:163, 1966.

Fraccaro, M.; Gerli, M.; Tiepolo, L.; Zara, C.: Analisi della variabilita cariotipica in un caso di neoplasia ovarica. *Minerva Ginec,* 18:187, 1966.

Fraccaro, M.; Mannini, A.; Tiepolo, L.; Gerli, M.; Zara, C.: Karyotypic clonal evolution in a cystic adenoma of the ovary. *Lancet,* 1:613, 1968.

Fritz-Niggli, H.: Analyse menschlicher Chromosomen. I. Karyotyp eines Mammakarzinoms. *Path Microbiol,* 17:340, 1954.

Gaffuri, S., and Bertoli, S.: Analisi cromosomica in alcuni tumori maligni dell'apparato genitale feminile. *Minerva Ginec,* 16:607, 1964.

Galton, M., and Benirschke, K.: Forty-six chromosomes in an ovarian teratoma. *Lancet,* 2:761, 1959.

Goodlin, R. C.: Karyotype analysis of gynecologic malignant tumors. *Am J Obstet Gynecol,* 84:493, 1962.

Greene, C. R.: The frequency of maldevelopment in man. *Am J Obstet Gynecol,* 90 (17):999, 1964.

Gropp, H.; Wolf, U.; Pera, F.: Chromatin und Chromosomenstatus beim Mammakarzinom. *Dtsch Med Wschr,* 90:637, 1965.

Gropp, H.; Pera, F.; Lohmann, H.; and Wolf, U.: Untersuchungen über die Anzahl der X-Chromosomen beim Mammakarzinom. *Z Krebsforsch,* 69:326, 1967.

Grouchy de, J.; Vallee, G.; Lamy, M.: Analyse chromosomique directe de deux tumeurs malignes. *C R Acad Sci,* 256:2046, 1963.

Hansen-Melander, F.; Kullander, S.; Melander, Y.: Chromosome analysis of a human ovarian cystocarcinoma in the ascites form. *J Nat Cancer Inst,* 16:1067, 1956.

Hughes, D. T.: The role of chromosomes in the characterization of human neoplasms. *Europ J Cancer,* 1:233, 1965.

Hughes, E. C., and Czermely, T. V.: Chromosome constitution of human endometrium. *Amer J Obstet Gynecol,* 93:777, 1965.

Hughes, E. C., and Csermely, T. V.: Chromosome constitution of human endometrium. *Nature,* 209:326, 1966.

Ishihara, T.; Moore, G. E.: Sandberg, A. A.: Chromosome constitution of cells in effusions of cancer patients. *J Nat Cancer Inst,* 27:893, 1961.

Ishihara, T., and Sandberg, A. A.: Chromosome constitution of diploid and pseudodiploid cells in effusions of cancer patients. *Cancer,* 16:885, 1963.

Ishihara, T.; Kikuchi, Y.; Sandberg, A. A.: Chromosomes of twenty cancer effusions. Correlation of karyotypic, clinical, and pathologic aspects. *J Nat Cancer Inst, 30*:1303, 1963.

Jackson, J. F.: Chromosome analysis of cells in effusions from cancer patients. *Cancer, 20*:537, 1967.

Jones, H. W., Jr.; Katayama, K. P.; Stafl, A.; Davis, H. J.: Chromosomes of cervical atypia, carcinoma in situ, and epidermoid carcinoma of the cervix. *Obstet Gynecol, 30*:790, 1967.

Jones, H. W., Jr.; Davis, H. J.; Frost, J. K.; Park, I.; Salimi, P.; Tseng, P.; Woodruff, J. D.: The value of the assay of chromosomes in the diagnosis of cervical neoplasia. *Am J Obstet Gynecol, 102*:624, 1968.

Kallenberger, A.: Geschlechts Chromatin bei Mammakarzinom. *Med Wschr, 94*:1450, 1964.

Kallenberger, A.; Hagmann, A.; Meier-Ruge, W.; Descoeudres, C.: Beziehungen zwischen Sexchromatin vorkommen, Kerngröbe, und DNS-werten in Mammatumoren und ihre Bedeutung für die Überlebenzeit. *Med Wschr, 97*:678 1967.

Katayama, K. P., and Jones, H. W., Jr.: Chromosomes of atypical (adenomatous) hyperplasia and carcinoma of the endometrium. *Am J Obstet Gynecol, 97*:978, 1967.

Katayama, K. P., and Masukawa, T.: Ring chromosomes in a breast cancer. *Acta Cytol, 12*:159, 1968.

Kawasaki, M.: Alteration of chromosomes in the peripheral leukocytes by administration of anti-neoplastic drugs to patients with gyneco-obstetric tumors. *J Jap Obstet Gynec Soc, 20*:413, 1968.

Kawasaki, M.: Peripheral leukocyte chromosome anomaly in cases of irradiation therapy after radical surgery of uterine cancer. *J Jap Obstet Gynec Soc, 20*:491, 1968.

Kawasaki, M.: Chromosomal studies of trophoblastic tumors. *J Jap Obstet Gynec Soc, 20*:699, 1968.

Kimel, V. M.: Clinical-cytological correlations of mammary carcinoma based upon sex chromatin counts. *Cancer, 10*:922, 1957.

Kirkland, J. A.: Atypical epithelial changes in the uterine cervix. *J Clin Path, 16*:150, 1963.

Kirkland, J. A.: Chromosomes in uterine cancer. *Lancet, 1*:152, 1966.

Kirkland, J. A.: Mitotic and chromosomal abnormalities in carcinoma in situ of the uterine cervix. *Acta Cytol, 10*:80, 1966.

Kirkland, J. A.: Chromosomal and mitotic abnormalities in preinvasive and invasive carcinomas of the cervix. *Aust N Z J Obstet Gynaec, 6*:35, 1966.

Kirkland, J. A., and Stanley, M. A.: The cytogenetics of carcinoma of the cervix. *Aust N Z J Obstet Gynaec, 7*:189, 1967.

Kirkland, J. A.; Stanley, M. A.; Cellier, K. M.: Comparative study of histologic and chromosomal abnormalities in cervical neoplasia. *Cancer, 20*:1934, 1967.

Klug, W.: Endocrine Therapie des Mammakarcinoms der Frau in Albangigkeit von der Zahl der Barrschen Zellkernkörper. *Chirurg, 40*:33, 1969.

Koller, P. C.: Abnormal mitosis in tumors. *Br J Cancer, 1*:38, 1947.

Levy, I. S., and Carel, R.: Variation in the incidence of sex chromatin: a reappraisal. *Acta Cytol, 12*:352, 1968.

Llewelly-Jones, D.: Trophoblastic tumors: geographical variations in incidence and possible aetiological factors. *J Obstet Gynaec Br Comm, 72*:242, 1965.

Loke, Y. W.: Sex chromatin of hydatiform moles. *J Med Genet, 6 (1)*:22, 1969.

Macklin, H. T.: Genetic considerations in human breast and gastric cancer. *Genetics and Cancer, 4*08; 1959.

Makino, S.; Tonomura, A.; Ishihara, T.: Studies on the chromosomes of some types of human cancer cells. *Zool Mag (Tokyo), 68*:142. 1959.

Makino, S.; Ishihara, T.; Tonomura, A.: Cytological studies of tumors, XXVII. The chromosomes of thirty human tumors. *Z Krebsforsch, 63*:184, 1959.

Makino, S.: Sasaki, H. S.; Fukuschima, T.: Preliminary notes on the chromosomes of human chorionic lesions. *Jap Acad Proc, 39*:54, 1963.

Makino, S.; Sasaki, H. S.; Fukushima, T.: Diploid chromosome constitution in human chorionic lesions. *Lancet, 2*:1273, 1964.

Makino, S.; Sasaki, H. S.; Tonomura, S.: Cytological studies of tumors-XL. Chromosome studies in 52 human tumours. *J Nat Cancer Inst, 32*:741, 1964.

Makino, S. et al.: Cytological studies of tumors, XLI. Chromosomal instability in human chorionic lesions. *Okajimas Fol Anat Jap, 40*:439, 1965.

Manna, G. K.: Chromosome number of human endometrium. *Nature, 176*: 354, 1955.

Manna, G. K.: Chromosome number of human cervix uteri. *Nature, 176*:355, 1955.

Manna, G. K.: A study on the chromosome number of human neoplastic uterine cervix tissue. *Proc Zool Soc (Calcutta), Mookerjee Memorial Vol.,* 95; 1957.

Manna, G. K.: The relative frequencies of different mitotic stages with some of their abnormalities in non-neoplastic and neoplastic human cervix uteri. *Proc Zool Soc* (Calcutta), *15*:1, 1962.

Marquez-Monter, H.: The superfemale mole. *Lancet, 2*:202, 1962.

Miles, C. P.: Chromosome analysis of solid tumors. II. Twenty-six epithelial tumors. *Cancer, 20*:1274, 1967.

Miles, C. P.: Chromosome analysis of solid tumors. I. Twenty-eight nonepithelial tumors. *Cancer, 20*:1253, 1967.

Moricard, R., and Cartier, R.: Chromosomal and cytoplasmic cytopathology of intraepithelial squamous cell epithelioma of the cervix uteri. *Ciba Foundation Study Group, 3*:28, 1959.

Ojima, Y.; Inui, N.; Makino, S.: Cytochemical studies on tumor cells. V.

Measurement of desoxyribonucleic acid (DNA) by Feulgen microspectrophotometry in some human uterine tumors. *Gann, 51*:371, 1960.

Polani, P. E.: Chromosome anomalies and abortions. *Develop Med Child Neurol, 8*:67, 1966.

Rashad, M. N.; Fathalla, M. F.; Kerr, M. G.: Sex chromatin and chromosome analysis of ovarian teratomas. *Am J Obstet Gynecol, 96*:461, 1966.

Rask-Madsen, J., and Philip, J.: The chromosome complement of human endometrium. *Cytogenetics, 9*:24, 1970.

Richards, B. M., and Atkin, N. B.: DNA content of human tumours: Change in uterine tumours during radiotherapy and their response to treatment. *Br J Cancer, 13*:788, 1959.

Richards, B. M., and Atkin, N. B.: The differences between normal and cancerous tissues with respect to the ratio of DNA content and chromosome number. *Acta Un Int Cancer (Brussels), 16*:124, 1960.

Richart, R. M., and Corfman, P.: Chromosome number and morphology of human preinvasive neoplasm. *Science, 144*:65, 1964.

Richart, R. M., and Wilbanks, G. D.: The chromosomes of human intraepithelial neoplasia. Report of 14 cases of cervical intraepithelial neoplasia and review. *Can Res, 26*:60, 1966.

Ruffié, J.; Marquès, P.; Mourlan, A. M.: Étude cytogénétique de deux tumeurs cancéreuses. *C R Acad Sci* (Paris), *258*:1935, 1964.

Sachs, L.: Subdiploid chromosome variation in man and other mammals. *Nature, 172*:205, 1953.

Sandberg, A. A.; Yamada, K.; Kikuchi, Y.; Takagi, N.: Chromosomes and causation of human cancer and leukemia. III. Karyotypes of cancerous effusions. *Cancer, 20*:1099, 1967.

Sasaki, M.; Fukushima, T.; Makino, S.: Some aspects of the chromosome constitution of hydatidiform moles and normal chorionic villi. *Gann, 53*:101, 1962.

Scholl, A.; Fischbach, H.; Morl, F.; Rickert, H.; Bohle, A.: Ergebnisse der Behandlung des Mammakarzinoms der Frau mit gegengeschlechtlichen Hormonen unter Berücksichtigung der Baarschen Zellkernkörpen. *Ver Deutsch Ges Path, 52*:426, 1968.

Serr, D. M.; Padeh, B.; Mashiach, S.; Shaki, R.: Chromosomal studies in tumors of embryonic origin. *Obstet Gynec, 33*:324, 1969.

Sharma, G. P.; Parshad, R.; Agnish, N. D.: Chromosome number in some malignant human tumors. *Res Bull Panjab Univ, 14*:99, 1963.

Sherman, A. I.: Chromosome constitution of endometrium. *Am J Obstet Gynec, 34*:753, 1969.

Shirley, R. L.: The nuclear sex of breast cancer. *Surg Gynec Obstet, 125*:737, 1967.

Silverberg, E., and Grant, R. N.: Cancer Statistics. 1970. *Ca—A Cancer J for Clinicians, 20*:10, 1970.

Šlot, E.: A karyologic study of the cancer of the ovary and the cancer cells in ascitic effusions. *Neoplasma, 14*:3, 1967.
Šlot, E. I., and Frauenklin, J. E.: The chromosome count in the stem cells of ovarian carcinoma and its relation to chemotherapeutic results. *Zbl Gynäk, 90*:210, 1968.
Spriggs, A. I.; Boddington, M. M.; Clarke, C. M.: Carcinoma in situ of cervix uteri: some cytogenetic observations. *Lancet, 1*:1383, 1962.
Spriggs, A. I., and Boddington, M. M.: Chromosomes of human cancer cells. *Br Med J, 2*:1431, 1962.
Spriggs, A. I.: Karyotype changes in human tumour cells. *Br J Radiol, 37*:210, 1964.
Stanley, M. A., and Kirkland, J. A.: Cytogenetic studies of endometrial carcinoma. *Am J Obstet Gynecol, 104*:1070, 1968.
Stolte, L. A. M.; Kessel, H. I. A.; Seelen, J. C.; Tijdink, G. A. J.: Chromosomes in hydatidiform moles. *Lancet, 2*:1144, 1960.
Tagliani, L.; Mastrangelo, C.; Curcio, S.: Studio citogenetico dei tumori utero-ovarici. Nota preventiva. *Arch Ostet Ginec, 68*:1, 1963.
Takemura, T.: Chromosome survey of normal human endometrium and endometrial carcinoma. *J Jap Obst Gynec Soc, 7*:300, 1960.
Timonen, S.: Mitosis in normal endometrium. *Acta Obstet Gynec Scand, 31*:1, 1950.
Tjio, J. H., and Puck, T. T.: The somatic chromosomes of man. *Proc Nat Acad Sci, 44*:1229, 1958.
Toews, H. A.; Katayama, K. P.; Jones, H. W.: Chromosomes of normal and neoplastic ovarian tissue. *Obstet Gynec, 32*:465, 1968.
Toews, H. A.; Katayama, K. P.; Masukawa, T.; Lewison, E. F.: Chromosomes of benign and malignant lesions of the breast. *Cancer, 22*:1296, 1968.
Tonomura, A.: A chromosome survey in 6 cases of human uterine cervix carcinoma. *Jap J Genet, 34*:401, 1959.
Tonomura, A.: Cytological studies of tumors—XXXII. Chromosome analysis in stomach and uterine carcinoma. *J Fac Sci Hokkaido Univ* (Ser VI Zool), *14*:149, 1959.
Tortora, M.: Chromosome analysis of four ovarian tumors. *Acta Cytol, 11*:225, 1967.
Tortora, M.: Chromosome studies in gynecological cancer. *Arch Ostet Ginec, 68*:437, 1963.
Tseng, P. Y., and Jones, H. W.: Chromosome constitution of carcinoma of the endometrium. *Obstet Gynecol, 33*:741, 1969.
Wakonig-Vaartaja, R.: A human tumor with identifiable cells as evidence for the mutation theory. *Br J Cancer, 16*:616, 1962.
Wakonig-Vaartaja, R.: Chromosomes in gynaecological malignant tumors. *Aust N Z J Obstet Gynec, 3*:170, 1963.
Wakonig-Vaartaja, R., and Hughes, D. T.: Chromosomal anomalies in dysplasia, carcinoma in situ and carcinoma of cervix uteri. *Lancet, 2*:756, 1965.

Wakonig-Vaartaja, R., and Kirkland, J.: A correlated chromosomal and histopathologic study of preinvasive lesions of the cervix. *Cancer, 18*:1101, 1965.

Wakonig-Vaartaja, R., and Hughes, D. T.: Chromosome studies in 36 gynaecological tumors: of the cervix, corpus uteri, ovary, vagina, and vulva. *Europ J Cancer, 3*:263, 1967.

Walker, B. E., and Boothroyd, E. R.: Chromosome numbers in somatic tissues of mouse and man. *Genetics, 39*:219, 1954.

Witkowski, R., and Zabel, H.: Chromosomenaberrationen der Tumorzellen aus dem Aszites bei Ovarial-Karzinom. *Acta Biol Med Germ, 16*:95, 1966.

WuMin: Chromosomes of cancer cells. *Vopr Onkol, 7*:9, 1961.

Yamada, K., and Sandberg, A. A.: Chronology and pattern of human chromosome replication. III. Autoradiographic studies on cells from cancer effusions. *J Nat Cancer Inst, 36*:1057, 1966.

Yamada, K.; Takagi, N.; and Sandberg, A. A.: Chromosomes and causation of human cancer and leukemia. II. Karyotypes of human solid tumors. *Cancer, 19*:1879, 1966.

Yamada, K., and Sandberg, A. A.: Karyotypes of 18 human solid tumors. Annual Meeting of the American Association for Cancer Research, 1965 (Abstract in program).

Chapter XIII

MALE REPRODUCTIVE SYSTEM

CHROMOSOMALLY analyzed neoplasms of the male genitals have been mainly testicular tumors. This group of malignancies constitutes only 1 percent of all human cancers and chiefly affects the 20 to 40 year-old group. Rarely do they occur in Negroes. They arise in undescended testes with relatively high frequency, abdominal testes being more often involved than inguinally retained testes. Malignant testicular tumors are often found in patients with male pseudohermaphroditism, especially in those cases with XO/XY mosaicism and a well-developed uterus. They have been noted in about 50 to 18 percent of patients with this syndrome (Jirásek, 1969, Campenhout et al., 1969).

For clarity, the classification proposed by Dixon and Moore (1953) and modified by Jirásek (1971) is presented.

A. Tumors originating from the epithelium of the sex cords or seminiferous tubules
  (1) dysgerminoma
    (a) *dysgerminoma* (pure seminoma, gonocytoma I)
    (b) *embryonic dysgerminoma* (embryonic carcinoma, dysgenic gonadoma, gonocytoma II)
    (c) *gonadoblastoma* (gonocytoma III)
  (2) malignant teratomas
    (a) *teratoblastoma* (teratocarcinoma)
    (b) *choriocarcinoma*
    (c) *teratoma with choriocarcinoma or with seminoma* (with dysgerminoma)
B. Tumors originating from the gonadal blastema and interstitium
  (a) *arrhenoblastoma*
  (b) *interstitial cells containing sarcoma*
  (c) *fibrosarcoma*

Atkin and Baker (1966) reported seminoma with chromosome counts ranging from 60 to 63 with a mode of 62 and distinct markers. Marker chromosomes resembled No. 1 and No. 3 in size and were seen in all four cells karyotyped. Twenty-three chromosomes were in group C, six in group D, and three in group G.

Atkin et al. (1966) measured DNA content and analyzed chromosomes in a seminoma having a mode of 108, and in a case of malignant teratoma of the testis with a mode of 53.

Galton et al. (1966) studied eight testicular teratomas. Modal numbers ranged between 53 and 111 and different markers were found in most metaphases. The majority of extra chromosomes belonged to groups C, E, F, and G. Most of the authors' attention was devoted to sex chromatin which was found in five of eight tumors.

Martineau (1966) attracted attention to large marker chromosomes, which she consistently found in four testicular tumors in a high proportion of cells. Case 1 was a seminoma, counts ranging from 63 to 74, with modes of 72 and 73. A long dicentric marker was found in 38 cells, a long acrocentric marker in 64 cells, and a long B type marker in 29 of 69 cells karyotyped. Case 2 was a seminoma and malignant teratoma of the right testis. Counts ranged from 45 to 124. A long marker was noted in 58 of 68 cells karyotyped. Most of the cells examined probably came from seminomatous parts of the tumor. Case 3, a malignant teratoma, had a range of 44 to 160 chromosomes per cell and a long dicentric marker in 28 of 34 cells. Case 4 was a malignant teratoma with a range of 41 to 125 chromosomes. A long marker, similar to that in three other tumors, was present in 35 of 43 examined metaphases. The similarity of morphology and size of the marker were remarkable as well as the fact that it never occurred in duplicate in any cell in all four tumors.

The data of Miles (1967) did not support the hypothesis of a specifc testicular tumor marker. A long chromosome having similar morphology to Martineau's marker was seen in only one of his series of five tumors, a metastatic pleomorphic seminoma with chromosome numbers ranging from 69 to 75. All aneuploid cells possessed a very long submetacentric chromosome nearly twice the length of a No. 1 homolog. Case 2, a well-differentiated seminoma with inflammatory infiltrate, yielded only metaphases with normal 46 chromosomes. Case 3 was a metastatic teratocarcinoma with mostly near-diploid cells and two karyotypes with 50 and 63 chromosomes. Cases 4 and 5 were metastatic embryonal carcinomas with normal diploid chromosomes. Examined cells were most likely nonmalignant because the specimens histologically showed considerably hyperplastic lymphoid tissue.

Rigby (1968) analyzed ten testicular tumors (see Table XXX). Karyotypes of all cells were diverse. Extra chromosomes were distributed irregularly among all Denver groups with the most prominent

TABLE XXX
CHROMOSOMES IN TESTICULAR TUMORS

Author	Mode	Long Marker (Martineau)	Tumor
Atkin and Baker (1966)	62	(?) +	Seminoma
Atkin et al. (1966)	108	?	Seminoma
Atkin et al. (1966)	53	?	Malignant teratoma
Martineau (1966)	73 and 72	+	Seminoma
Martineau (1966)	?	+	Seminoma and malignant teratoma
Martineau (1966)	?	+	Malignant teratoma
Martineau (1966)	?	+	Malignant teratoma
Miles (1967)	?	+	Seminoma
Miles (1967)	Normal 46 (?)	—	Seminoma
Miles (1967)	?	—	Malignant teratoma
Miles (1967)	Normal 46 (?)	—	Embryonal carcinoma
Miles (1967)	Normal 46 (?)	—	Embryonal carcinoma
Rigby (1968)	68	+	Seminoma
Rigby (1968)	76	+	Seminoma
Rigby (1968)	61	+	Seminoma
Rigby (1968)	77	—	Seminoma
Rigby (1968)	73	+	Seminoma and malignant teratoma
Rigby (1968)	69	+	Seminoma
Rigby (1968)	52	—	Malignant teratoma
Rigby (1968)	58	—	Malignant teratoma
Rigby (1968)	58	—	Malignant teratoma
Fischer and Golob (1967)	54-56	+	Seminoma

increase being in group C. An outstanding feature of the tumors was the presence of markers. A chromosome with a subterminal centromere was present in every tumor in 35 percent to 95 percent of cells (10 metaphases were karyotyped for each tumor). A long marker with a secondary centromere (or constriction) was found in at least 50 percent of cells in his Cases 1, 2, 6, and 7.

Apparently the specificity of a long marker described by Martineau (1966) is the most interesting and controversial finding. If there is specificity of this marker, it would be site specific because this long chromosome has been observed in testicular tumors of different histology and etiopathology as in seminoma and malignant teratoma. The morphology of the marker was similar but not identical in different reported specimens. The proportion of long and short arms as well as the relative length of the chromosome varied. On the other hand, the marker has not been seen in duplicate in any cell and appears to be quite characteristic for stemline cells in the tumors described. In one case, it occurred in seminomas of both testes of a patient (Rigby, 1968).

Due to the lack of more convincing data, the question of Martineau's marker cannot be answered at present and the proof of its specificity remains to be established.

This chapter should include two cases of squamous cell carcinoma of the penis reported by Tabata (1959). Both cases had stemline in the range 71 to 80 (47 percent). No abnormal marker is reported. Forty-three cells were counted.

REFERENCES

Atkin, N. B.; Mattinson, G.; Baker, M. C.: A comparison of the DNA content and chromosome number of fifty human tumours. *Br J Cancer,* 20:87, 1966.

Atkin, N. B., and Baker, M. C.: Chromosome abnormalities as primary events in human malignant disease: evidence from marker chromosomes. *J Nat Cancer Inst,* 36:539, 1966.

Atkin, N. B., and Baker, M. C.: Possible differences between the karyotypes of preinvasive lesions and malignant tumors. *Br J Cancer,* 23:329, 1969.

Atkin, N. B.: Cytogenetic studies on human tumors and premalignant lesions: The emergence of aneuploid cell lines and their relationship to the process of malignant transformation in man. 23rd Annual Symposium on Fundamental Cancer Research, Houston, Texas, 1969.

Campenhout, J. van; Lord, J.; Vauclair, R.; Lanthier, A.; Bernard, M.: The phenotype and gonadal histology in XO/XY mozaic individuals: Report of two personal cases. *J Obstet Gynec Brit Cwlth 76*:631, 1969.

Fischer, P., and Golob, E.: Similar marker chromosomes in testicular tumours. *Lancet, 1*:216, 1967.

Galton, M.; Benirschke, K.; Baker, M. C.; and Atkin, N. B.: Chromosomes of testicular teratomas. *Cytogenetics, 5*:261, 1966.

Jirásek, J.: Personal communication, 1969.

Jirásek, J.: *Development of the Genital Apparatus and Male Pseudohermaphroditism.* Johns Hopkins Press, Baltimore, 1971.

Martineau, M.: A similar marker chromosome in testicular tumors. *Lancet, 1*:839, 1966.

Miles, C. P.: Chromosome analysis of solid tumors. I. Twenty-eight non-epithelial tumors. *Cancer, 20*:1253, 1967.

Pierce, G. B. Jr.: The pathogenesis of testicular tumors. *J Urol, 88*:573, 1962.

Rigby, C. C.: Chromosome studies in ten testicular tumors. *Br J Cancer, 22*:480, 1968.

Tabata, T.: Karyological studies of human tumors. *Wakayama Med Soc J, 34*:963, 1959.

Chapter XIV

ALIMENTARY TRACT

PREMALIGNANT LESIONS

ALL CHROMOSOMAL INVESTIGATIONS of precanceroses of the alimentary tract have been carried out on colonic adenomas or polyps (papillary or villous adenomas, sessile or pedunculated adenomas). These lesions are regarded as the most important "etiologic" factor in colonic cancer and, in most cases, are of genetic origin.

Using the technique of microspectrophotometry on Fuelgen-stained nuclei, Stich *et al.* (1960) attempted to estimate the ploidy from determination of DNA content. Specimens from eight patients with rectal polyps yielded diploid DNA content in six and a predominant tetraploid DNA content in the other two. Four additional samples from one patient with hereditary multiple polyposis were examined and the DNA content was found to be in the diploid range in all of them.

It should be noted, however, that microspectrophotometry is not precise enough to establish the exact number of chromosomes and thus estimation of the diploid value of the specimens noted above may represent near-diploidy with one or more chromosomes supernumerary or absent, or pseudodiploidy with 46 chromosomes but aberrant chromosome morphology or markers.

Three years later, Lubs and Clark (1963) carried out exact chromosome counts on two karyotypes of benign adenomatous polyps removed by sigmoidoscope from a 68-year-old man. The chromosome spreads were obtained by direct method and yielded two karyotypes, both apparently normal diploid. An incidental sigmoid polyp with marked villous and adenomatous components was obtained from another patient. The chromosomes of twelve cells were counted and all except two were abnormal in number. Three of four karyotypes contained 47 chromosomes with supernumerary ones in the F group.

The most extensive study of the problem of colonic precancerosis was undertaken by Enterline and Arvan (1967). The authors classified their specimens into five groups as follows: (a) *adenoma without atypia*, (b) *adenoma with minor atypia*, (c) *adenoma with major atypia*, (d) *adenoma with villous growth pattern*, and (e) *adenocarcinoma without associated adenomatous elements*.

Chromosomal aberrations, mainly hyperploidy, were found with significant frequency in all groups, including six patients from the first group of typical "benign" adenoma. Two patients in the first group displayed a sharp mode of 47 chromosomes. One had a supernumerary chromosome in the C group, while the second had a supernumerary D group chromosome.

Examination of polyps with major and minor atypia (groups 2 and 3) revealed higher ploidy than those from group 1. In two cases with major atypia, the chromosome number ranged between 51 and 60 or 51 and 90, respectively. In polyps with less marked atypia, the count was lower: 48 to 50. There was a suggestion of stemlines in three of five cases and several large markers were noted. All groups contained pseudodiploid cells. In adenocarcinoma, hyperploidy was even more marked than in the other four groups.

This indirect suggestion that higher ploidy is associated with a more advanced degree of malignancy does not appear to fit the available data on frank carcinoma (see following section on carcinoma of colon, stomach, liver, etc.). In fact, Enterline and Arvan included in their study two cases of poorly differentiated carcinomas with diploid and near-diploid counts. In these cases, the authors suggested the possibility of selection against severe hyperploidy.

Examination of inflammatory tissue from the colon was performed as well and 47 metaphases were counted. As expected, all were normal diploid except for one cell with 48 chromosomes. Miles (1967) analyzed one adenomatous polyp of the colon but was able to construct only one karyotype. This karyotype had a normal diploid set of chromosomes. A second specimen—villous adenoma with *in situ* carcinoma—contained at least one cell with 48 chromosomes.

Two *villous adenomas* of the colon were analyzed by Kotler and Lubs (1967) and Lubs and Kotler (1967). As in the cases in group 4 of the Enterline and Arvan classifications, aberrant chromosome numbers were found in one specimen, ranging between 23 and 85. No predominant cell lines were observed and no common karyotypic alterations were found. Karyotypes showed groups C, F, and G to be chiefly affected. The other adenoma contained no markers and no aberrant stemline; chromsome counts ranged from 29 to 68.

Messinetti et al. (1968) karyotyped 18 cells in two cases of *benign adenomatous polyp* of the sigmoid colon without atypia. All cells had a normal diploid set. However, one specimen of adenomatous sessile polyp of the stomach had clearly aberrant chromosomes with a mode of 58 and chromosome number ranging between 45 and 106.

Abnormal chromosome findings were confirmed again by Atkin and Baker (1969) who examined a single *colonic polyp* having C-trisomy as the only change, two *rectal polyps* with C-trisomy, one rectal polyp with 49 chromosomes, and one rectal polyp having pseudodiploid constitution. The karyotype in the last case showed two markers, one being a ring chromosome, C-trisomy and missing chromosomes in groups A, B, and E.

Assuming that an abnormal karyotype is characteristic of the majority of cases of cancer, the data cited above on polyps and adenomas of the colon and rectum would then agree with the commonly held opinion of clinicians that these lesions are the precursors of colonic cancer. Malignant transformation of villous adenoma, for example, has been estimated to occur in 30 to 75 percent of cases. It may be that even this estimation is not sufficiently high since in familial polyposis almost all affected individuals may be expected to develop carcinoma of the colon 10 to 15 years following its initial appearance.

Chromosomal analysis of adenomas without signs of atypia (Enterline and Arvan, 1967) justifies the hypothetical developmental sequence of adenomatous polyp→adenoma with atypia→adenocarcinoma.

ORAL CAVITY, ESOPHAGUS

Reports of chromosomes obtained from oral tumors are sparse, despite their relatively high frequency and great histopathological variety. No odontogenic tumors have been examined. Most reports have been concerned with oral squamous cell carcinomas.

One of the earliest reports on chromosomes in human cancer was made by Hsu in 1965. In addition to other tumors in his series, he examined a squamous cell carcinoma of the submandibular gland. Cells were incubated for six days and 200 divisions examined. In none could exact counts be made, but marked polyploidy and fragmentation were visible.

Atkin and Richards (1956) used microspectrophotometry on Feulgen-stained nuclei to estimate the DNA content of cells from squamous cell carcinomas of the tongue and buccal mucosa. In both tumors, considerable variation of DNA content was found suggesting aneuploidy.

The only reports on exact chromosome number in tumors of the oral cavity are those of Makino et al. (1959) and von Ertl et al. (1970). Makino examined four neoplasms of the maxilla, three squamous cell carcinomas and one not identified histologically. Karyotypes were not constructed but chromosome counts showed a hypertriploid modal number in three tumors while one tumor exhibited no predominant stemline. One tumor had a mode of 71 (33.3% cells), the second a mode of 73 (38.4% cells), while the third had a mode of 69 (38.4% cells). In the fourth specimen, counts ranged widely between 63 and 104 chromosomes.

Von Ertl et al. (1970) cultured three carcinomas of the oral cavity for 48 hours and in all obtained wide spread counts. No notable markers and no prominent stemlines were observed.

White and Cox (1967) reported a highly cellular rhabdomyosarcoma of the temporal muscle of a 3-year-old girl. Chromosome spreads were obtained by direct method in two recurrences, four months apart. In the first sampling, chromosome counts showed no clear mode and were spread over a range of 68 to 72. In a second recurrence, there was a definite accumulation of counts at 67 and 68. Culturing the cells, the authors observed a clearly downward shift of modal numbers: after one month the mode was 66; after four months, 64; after 12 months, 57 to 59; and after 18 months, 56 and 57. Persistence of markers was observed as well as double chromatin bodies from cells of the first recurrence but not of the second.

The only report on squamous cell carcinoma of the esophagus is that of Lubs and Clark (1963). In six cells the chromosomal number was 43 or less. Only one karyotype was constructed and contained a large acrocentric marker as well as a very long marker similar to an A or B group chromosome, probably representing associations of missing acrocentrics.

There is not enough available data to attempt the association of chromosomal findings either with site or histopathologic specificity of oral tumors. No data have been published on specific oral pre-canceroses.

STOMACH

Hsu (1954), Levan (1956), Atkin and Richards (1956), and Ising and Levan (1957), reported chromosome studies and DNA content estimation of gastric carcinoma, carcinoma of the colon, and carcinoma of the anus. Exact chromosome counts in addition to microspectrophotometry were made only by Ising and Levan who found a definite stemline of 82 chromosomes and a consistently present ring marker in two samples obtained two weeks apart from the same patient. Some cells from the gastric carcinoma contained as many as 220 chromosomes.

The first large scale chromosomal study of gastric carcinoma was reported by Makino et al. (1959), who examined ten cases among their series of 30 malignant tumors. Three specimens were obtained from solid tumors and seven from peritoneal effusions. One solid tumor had a mode of 56. Two others did not have enough mitoses to establish the modal number, counts ranging from 89 to 112 and 80 to 109, respectively. In five effusion samples, the modal numbers were as follows: 42; 45; 82 to 89; 72 and 76; 74.

In one patient, three effusion samples were obtained during a six-week-period. Prior to the first sampling, the patient was treated by x-irradiation. The first sampling showed a predominant stemline of 45 chromosomes while some cells had 42. The last sampling revealed a predominant stemline of 42 suggesting that cells with 42 chromosomes gradually prevailed, possibly due to their better adaptivity to the altered environment caused by radiation.

The evolution of stemlines in gastric carcinoma was further studied by Ishihara (1959) and Tonomura (1960). Both authors observed a diminution in the number of cells with extreme hyperploidy. This was true for cells without apparent relation to existing stemlines and for those which most likely originated from stem cells by endoreduplication. Ishihara (1959) studied three samples from gastric carcinoma effusions taken over a period of three months. The patient was irradiated prior to and during the sampling. The tumor had three main cell populations: two near-diploid and one in the range of 80 to 84 chromosomes. Analogous development of stemlines was observed as in the case described by Makino (1959). Cells with 45 chromosomes were of highest frequency in the first sampling (28%) and of lower frequency in the second and third samplings. Stemline cells with 42 chromosomes were lower in the first sampling and their occurrence markedly increased in the second (34%) and third (31.2%) samplings. Cells with chromosome numbers 80 to 84, which probably arose by endoreduplication from stem cells containing 42 chromosomes, gradually decreased in number.

Tonomura (1960), following similar lines of investigation, examined cancerous effusion from "carcinoma simplex of the stomach with metastases." Samples were drawn on four occasions during the period of one month (Table XXXI). The patient was treated with Carzinophilin (tumor inhibiting substance produced by Streptomyces) before and after the first sampling but not after the second sampling. Each examination showed a marked peak of cells having 62 chromosomes which persisted without being affected by treatment and even increased in the 3rd and 4th samples. On the other hand, very high-ploid cells, (octoploid) as well as cells with chromosomal breaks, were greatly diminished.

Stich and Steele (1962), using the microspectrophotometric two

TABLE XXXI
FREQUENCY DISTRIBUTIONS OF TUMOR CELLS BELOW AND ABOVE TETRAPLOIDY IN FOUR SAMPLINGS

Sampling	Cells 4n (92) or Less	Cells More than 4n (92)	Total Cells Counted
1.	39 (42.9%)	52 (57.1%)	91
2.	16 (59.3%)	11 (40.7%)	27
3.	29 (76.3%)	9 (23.7%)	38
4.	34 (100%)	0 (0%)	34
Total	118	72	190

Note: From Tonomura, 1960.

wavelength method of Patau, examined adenocarcinoma of the large bowel, leiomyosarcoma of the colon and adenocarcinoma of the stomach. The gastric tumor showed that only cells with hypertetraploid DNA content of the nondividing cell population displayed a mosaic, inconsistent, composition.

The leiomyosarcoma cells were of two modes: hypotetraploid and hypertetraploid.

Makino et al. (1964) published the most extensive study on gastric carcinoma, examining 30 tumor effusions. In most, clear stemlines could be detected despite a rather remarkable spread of chromosome numbers. Four tumors contained an essentially normal karyotype with 46 chromosomes. Two tumors were examined before and after treatment with Toyomycin, Carzinophilin and Au^{198}. In both, therapy failed to change the chromosome constitution of the stem cells, which contained 51 chromosomes in the first tumor and 63 and 65 in the second tumor.

Several other authors contributed to the problem of gastric cancer (Table XXXII), confirming the absence of typical markers or consistency in cell line development.

The above data on chromosomes from gastric carcinoma illustrate the great variability of chromosome numbers and also the variability of chromosome numbers of stemline cells, which ranged from 42 to 120. Studies of possible evolution of stemlines within a single tumor

TABLE XXXII
CHROMOSOMES OF CARCINOMA OF STOMACH

Author	Stemline	Markers	Site
Tonomura (1959)	72 and 76	—	effusion
Awano and Tuda (1959)	45	—	effusion
	49	—	effusion
Ishihara et al. (1961)	64	—	effusion
	56	—	effusion
	46 pseudo	—	effusion
	90	—	effusion
Spriggs et al. (1962)	60 and 120	—	effusion
	—	—	effusion
Ishihara and Sandberg (1963)	71	—	solid
Ishihara et al. (1963)	—	large acrocentric	effusion
Sandberg et al. (1967)	72	—	effusion
Miles (1967)	—	large metacentric	solid
Messinetti et al. (1968)	48	2 metacentric	solid

did not yield convincing results. Two reports presented data showing one stemline gaining advantage over the other. However, two later studies did not confirm this observation: Makino (1959) and Ishihara (1960) noted stemlines with hypodiploid numbers of 42 chromosomes being replaced by stemlines with a higher number of chromosomes, while Tonomura (1960) and Makino (1964) reported stemlines with hyperdiploid numbers 62, 51, 63, and 65 remaining unchanged, apparently not being affected by treatment. The question then remains unanswered whether hyperdiploid cell lines in effusions of gastric cancer are generally more stable and adaptable than cells with hypodiploid nuclei.

There is no sign of a specific marker being common to gastric cancer, though some acrocentric markers were suggested to play an important role in the development of malignancy.

Representation of different chromosomal groups in malignant karyotypes was investigated by van Steenis (1966). The author analyzed karyotypes from 14 gastric carcinomas reported by Makino (1964) and from 12 tumors published by Ishihara (1963). After excluding the structurally abnormal chromosomes from his calculations, the distribution of chromosomes over the Denver classification groups was expressed as a percentage of the total number of normal chromosomes per cell. Group C contained significantly higher numbers of supernumerary chromosomes and groups D and G had significantly fewer chromosomes than expected. These calculations were of a different nature than the data of Atkin and Baker (1969) (see following section). Van Steenis assumed that the excess of C group chromosomes was most likely caused by chromosomes formed by a centric fusion type of translocation between acrocentrics, which he found consistent with the finding of underrepresentation of acrocentric groups D and G. Van Steenis' hypothesis was, however, challenged by Sandberg et al. (1968), on the basis of vector analysis of acute leukemias and solid cancers.

COLON, RECTUM, ANUS

Among gastrointestinal neoplasms, tumors of the colon are second in frequency to carcinoma of the stomach. The most important etiologic factor is the presence of polyps, or more properly, adenomas (see section entitled "Premalignant Lesions" in this chapter) which frequently develop into typical adenocarcinoma. Tumors of the anus are usually epidermoid carcinomas, as a rule, poorly differentiated. They are relatively uncommon, compared with adenocarcinoma of the colon.

The first report on chromosomes of carcinoma of the rectum and "alimentary canal" was published by Koller (1947) when 48 chromosomes were considered to be the normal human diploid number. The author presented several photographs of tumor chromosomes, some being clearly polyploid, some hypodiploid.

Sandberg et al. (1963) found an unusually high ploidy in peritoneal effusion cells from anaplastic carcinoma of the sigmoid colon. In direct preparation, 51.5 percent of the cells contained over 600

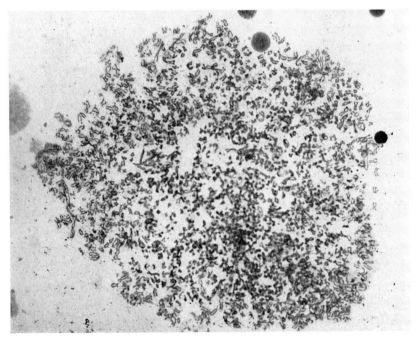

Figure 20. Metaphase spread from one cell in the peritoneal effusion of a patient with cancer of the colon. The number of chromosomes exceeds 1000, multiple markers could be detected. (Courtesy of Dr. A. A. Sandberg et al., Cancer, 16:1246, 1963.)

chromosomes, including chromosomes of abnormal size and rings. Some cells had over 1,000 chromosomes—even an estimation of 2,000 was made (see Fig. 20). A mode of 77 chromosomes was found in 21.5 percent of cells, suggesting that the original tumor had not been extremely hyperploid. The high numbers seemed to arise by endoreduplication since, in some cells, the topographic proximity of morphologically identical chromosomes was still observed. After culturing the cells for 12 days the frequency of markers did not change but highploidy cells disappeared. Only 8 percent of cells had over 600 chromosomes and cells with 77 chromosomes occurred in 42 percent.

Numerous other chromosomal analyses of carcinoma of the colon and rectum are summarized in Table XXXIII since no reason for detailed discussion does not seem merited.

In colonic carcinoma, chromosome numbers have ranged widely, being as low as 30 and reaching extreme hyperploidy of about 1000 to 2000 chromosomes. Nevertheless, the tendency to eliminate very high-ploidy cells in culture was observed. It appeared likely that cells with too many chromosomes are probably not capable of dividing under artificial conditions or that reversion to more normal mitotic division has occurred in culture (Sandberg et al., 1963). As in tumors from most other sites, correlation between the degree of malignancy and specific chromosomal constitution has not been clearly demonstrated. However, hyperdiploidy and near-triploidy seem to prevail in advanced tumors of the colon.

Large marker chromosomes occur quite frequently and, in some tumors, their incidence was reported in 100 percent of tumor cells. Often, they survive the process of evolution of stemlines and appear to be a characteristic feature of most of the tumor cells despite their difference in chromosome numbers. Atkin and Baker (1966) concluded that the consistent perpetuation of marker chromosomes strongly agrees with the hypothesis that human malignant tumors originate chiefly from a single cell in which chromosome changes have occurred (see General Part). This applies in the case of colonic cancer arising simultaneously from several polyps in familial multiple polyposis, i.e. the invasive tumor may develop in several places independently.

Interesting are the recent reports of Atkin and Baker (1969) and

TABLE XXXIII
CHROMOSOMES OF CARCINOMA OF COLON, RECTUM, AND ANUS

Author	Stemlines	Markers	Site
Richards and Atkin (1960)	60	—	rectum
	40	—	rectum
Ishihara et al. (1961)	46 and 45	large acrocentric in all cells	colon, effusion
Spriggs et al. (1962)	78	—	colon, effusion
Ishihara and Sandberg (1963)	65	—	colon, effusion
	—	3 markers in 90% cells	colon, colon effusion
Lubs and Clark (1963)	92	—	colon
	77	2 large acrocentric	colon
	—	—	colon
Yamada et al. (1966)	46 pseudo	various markers	colon
	46 pseudo	various markers	colon
	48	various markers	colon
	49	various markers	colon
	51	various markers	colon
	66 and 74	various markers	colon
	67 and 69	various markers	colon
Atkin and Baker (1966)	—	large metacentric	colon
	51-5	—	colon
	43-44	large metacentric	rectum
Jackson (1967)	63	—	colon, effusion
Miles (1967)	70-75	long metacentric	colon
	44	submetacentric	colon
	43, 80-81	—	colon
	83-88	—	colon
	80-83	—	colon
	70-76	—	colon
	85-86	—	rectum
	62-77	—	rectum
	65-80	—	rectum
Sandberg et al. (1967)	70-80	various markers	rectum, effusion
Messinetti et al. (1968)	60	metacentric	colon
	78	metacentric	colon
Lubs and Kotler (1967)	65-64		colon
Atkin and Baker (1969)	—	missing B, G, D,	colon
	—	missing B, G, D,	colon
	—	missing B, G, D,	colon
	—	missing B, G,	colon
	—	missing B, G,	rectum
	—	missing B, G, D,	rectum
	—	missing B, G, D,	rectum

Atkin (1969). Examining 23 malignant tumors, including seven carcinomas of the colon and rectum, the authors found significant underrepresentation of chromosomes of the B, D, and G groups in most of the tumors. Groups D and G taken together were deficient in all but one tumor. These data do not appear to be in agreement with the findings of Gofman et al. (1967) and Minkler et al. (1971) for tumors of the alimentary tract as well as of other sites. These latter authors calculated the "specific common chromosomal pathway for the origin of malignancy" to be an imbalance in the E_{16} chromosome, i.e. an excess of E_{16}, either absolute or relative to the other groups.

LIVER, PANCREAS, PERITONEUM

Despite the fact that primary hepatic carcinoma is very rare in Caucasians, the chromosomes of effusion cells from four primary tumors have been analyzed.

The first report on human liver tumor is that of Ishihara and Sandberg (1963). An effusion of liver carcinoma contained 30 percent cells with a normal set of 46 chromosomes. Because histological examination did not show any neoplastic cells, the authors suspected that they had examined only the nonmalignant host cells. Abnormal findings were reported by Makino et al. (1964) for another effusion of liver carcinoma (epithelium-like carcinoma in the ascites phase). Chromosome numbers ranged from 65 to 83 with a sharp mode of 75 chromosomes. Striking was an increased number (up to 16) of G group chromosomes.

Jackson (1967) examined an effusion of liver carcinoma with chromosome numbers between 41 and 103. Forty percent of the cells contained 46 chromosomes.

Sandberg et al. (1967) investigated peritoneal effusion of a liver carcinoma noting a flat mode of 79 chromosomes with three near constant markers. Chromosome A_1 was consistently diploid in all karyotypes.

Of interest is de Grouchy's (1969) analysis of a liver metastasis from a gastric carcinoma which has been kept in culture since 1962. Chromosomes of cultured tumor cells were examined in 1966 and again in 1968. A change in the mode from 64 chromosomes (1966) to 69 chromosomes (1968) was observed. Specific markers were retained unchanged. A loss in D group and a gain in C group chromosomes were ascertained.

There has been but a single report on effusion of carcinoma of the pancreas. Sandberg et al. (1967) found a modal number exceeding 100 chromosomes and a rather infrequent ring marker.

Peritoneal effusion of rare sarcoma of the mesentery was analyzed by Ishihara and Sandberg (1963). Two samplings were taken. In the first, all examined cells contained 46, apparently normal chromosomes. The second sampling still showed 88.8 percent normal diploid cells. Histological examination revealed an absence of malignant cells in the fluid.

Sandberg and Yamada (1966) elaborated on findings from effusion of cancer of the peritoneum. Fluid was withdrawn on four occasions within a single month. Three samples were obtained before and one after 5-fluorouracil therapy. Despite the fact that 65 percent of the cells were karyotyped, none of which was completely identical with any other, selection of one type of cell over another was observed. Some characteristics were constant such as two markers (98% cells) and the presence of 4 chromosomes in groups B and G (80% cells). The modal number ranged from 54 to 56. These features, in the authors' opinion, indicated maintenance of the basic chromosome pattern, which might be important for survival of the cancerous cells.

Chromosomal preparations of a solid retroperitoneal leiomyosarcoma were analyzed by Miles (1967). Three karyotypes exhibited only normal 46/XY cells despite histological examination which showed a very cellular tumor with a high mitotic rate and no apparent admixture of lymphocytes or other inflammatory components.

No specific chromosomal features of human hepatic, pancreatic, or peritoneal carcinoma were revealed. The findings of Miles (1967) require collection of more extensive data to be adequately explained.

REFERENCES

Atkin, N. B., and Richards, B. M.: Deoxyribonucleic acid in human tumours as measured by microspectrophotometry of Feulgen stain. A comparison of tumors arising at different sites. Br J Cancer, 10:769, 1956.

Atkin, N. B.; Mattinson, G.; Baker, M. C.: A comparison of the DNA content and chromosome number of fifty human tumors. Br J Cancer, 20:87, 1966.

Atkin, N. B., and Baker, M. C.: Chromosome abnormalities as primary events in human malignant disease: evidence from marker chromosomes. *J Nat Cancer Inst, 36*:539, 1966.

Atkin, N. B., and Baker, M. C.: Possible differences between the karyotypes of preinvasive lesions and malignant tumors. *Br J Cancer, 23*:329, 1969.

Atkin, N. B.: Cytogenetic studies on human tumors and premalignant lesions: the emergence of aneuploid cell lines and their relationship to the process of malignant transformation in man. Genetic Concepts and Neoplasia; 23rd Annual Symposium on Fundamental Cancer Research. The University of Texas, Houston, 1969.

Awano, T., and Tuda, E.: The chromosomes of stomach cancer and myelogeneous leukemias in comparison with normal complex. *Jap J Genet, 34*:220, 1959.

Enterline, H. T., and Arvan, D. A.: Chromosome constitution of adenoma and adenocarcinoma of the colon. *Cancer, 20*:1746, 1967.

Ertl, H. von; Schlegel, D.; Wieser, O.: Zytogenetische Untersuchengen an autochtonen, menschlichen Tumoren der Mundhöhle. *Deutsch Zahnaerztl Z, 25*:407, 1970.

Grouchy de, J.: Analyse chromosomique d'une tumeur cancéreuse humaine en culture organotypique. *Europ J Cancer, 5*:159, 1969.

Gofman, J. W.: Minkler, J. L.; Tandy, R. K.: A specific common chromosomal pathway for the origin of human malignancy. U S Dept Commerce Natl Bureau of Standards, Springfield, Virg, 22151. Publication No: UCRL 50356, 1967.

Hsu, T. C.: Mammalian chromosomes *in vitro*. IV. Some human neoplasms. *J Nat Cancer Inst, 14*:905, 1954.

Inui, N.: Histological and chromosomal studies in two human gastric carcinomas and in their metastatic lesions. *Jap J Genet, 41*:115, 1966.

Ishihara, T.: Cytological studies of tumors. XXXI—A chromosome study in a human gastric carcinoma. *Gann, 50*:403, 1959.

Ishihara, T.; Moore, G. E.; Sandberg, A. A.: Chromosome constitution of cells in effusions of cancer patients. *J Nat Cancer Inst, 27*:893, 1961.

Ishihara, T., and Sandberg, A. A.: Chromosome constitution of diploid and pseudodiploid cells in effusions of cancer patients. *Cancer, 16*:885, 1963.

Ishihara, T.; Kikuchi, Y.; Sandberg, A. A.: Chromosomes of twenty cancer effusions—correlation of karyotypic, clinical and pathologic aspects. *J Nat Cancer Inst, 30*:1303, 1963.

Ising, U., and Levan, A.: The chromosomes of two highly malignant human tumors. *Acta Path Microb Scand, 40*:13, 1957.

Jackson, J. F.: Chromosome analysis of cells in effusions from cancer patients. *Cancer, 20*:537, 1967.

Koller, P. C.: Abnormal mitosis in tumors. *Br J Cancer, 1*:38, 1947.

Kotler, S., and Lubs, H. A.: Comparison of direct and short-term tissue culture technics in determining solid tumor karyotypes. *Cancer Res, 27*:1861, 1967.

Levan, A.: Self-perpetuating ring chromosome in two human tumors. *Hereditas*, 42:366,1956.

Lubs, H. A., and Clark, R.: The chromosome complement of human solid tumors. *N Eng J Med*, 268:907, 1963.

Lubs, H. A., and Kotler, S.: The prognostic significance of chromosome abnormalities in colon tumors. *Ann Intern Med*, 67:328, 1967.

Macklin, M. T.: Genetic considerations in human breast and gastric cancer. *Genetics and Cancer*, Austin, Texas, Univ. Texas Press, 1959, p. 408.

Maeda, M.; Tabata, T.; Kusayama, S.; Kimura, T.: Chromosomes of human tumors. V. Chromosomal alteration in a metastatic lesion of a maxillary cancer. *Wakayama Med Rep*, 9:225, 1965.

Makino, S.; Ishihara, T.; Tonomura, A.: Cytological studies of tumors. XXVII. The chromosomes of thirty human tumors. *Z Krebsforsch*, 63: 184, 1959.

Makino, S.; Tonomura, A.; Ishihara, T.: Studies on the chromosomes of some types of human cancer cells. *Dobkutsaku Zasshi* (Tokyo), 68:142, 1959.

McConnell, T. S., Parsons, L.: Chromosome evaluation in familial polyposis of the colon. *Rocky Mountain Med J*, 65:51, 1968.

Messinetti, S.; Zelli, G. P.; Marcellino, L. R.; Tumino, G.: L'analisi cromosomica nella citodiagnostica dei versamenti ascitici di carcinomi gastroenterici. *Ann Ital Chir*, 42:800, 1965.

Messinetti, S.; Zelli, G. P.; Marcellino, L. R.; Tumino, G.: L'analisi cromosomica nello studio e nella diagnostica dei tumori epiteliali benigni e maligni del tubo gastroenterico. *Ann Ital Chir*, 42:817, 1965.

Messinetti, S.; Zelli, G. P.; Marcellino, L. R.; Alcini, E.: Benign and malignant epithelial tumors of the gastroenteric tract. Chromosome analysis in study and diagnosis. *Cancer*, 21:1000, 1968.

Miles, C. P.: Chromosome analysis of solid tumors. II. Twenty-six epithelial tumors. *Cancer*, 20:1274, 1967.

Minkler, J. L.; Gofman, J. W.; Tandy, R. K.: A specific common chromosomal pathway for the origin of human malignancy—II. *Br J Cancer*, 24:726, 1971.

Richards, B. M., and Atkin, N. B.: The difference between normal and cancerous tissues with respect to the ratio of DNA content to chromosome number. *Acta Un Int Cancer*, 16:124, 1960.

Sandberg, A. A.; Ishihara, T.; Moore, G. E.; Pickren, J. W.: Unusually high polyploidy in a human cancer. *Cancer*, 16:1246, 1963.

Sandberg, A. A., and Yamada, K.: Chromosomes and causation of human cancer and leukemia. I. Karyotype diversity in a single cancer. *Cancer*, 19:1869, 1966.

Sandberg, A. A.; Yamada, V.; Kikuchi, Y.; Takagi, N.: Chromosomes and causation of human cancer. III. Karyotypes of cancerous effusions. *Cancer*, 20:1099, 1967.

Sandberg, A. A.; Bross, I. D. J.; Takagi, N.; Schmidt, M. L.: Chromosomes and causation of human cancer and leukemia. IV. Vectorial analysis. *Cancer, 21*:77, 1968.

Sasaki, M. S.: Cytological effect of chemicals on tumors. XII. A chromosome study in a human gastric tumor following radioactive colloid gold (Au^{198}) treatment. *J Fac Sci Hokkaido Univ,* Ser VI 2001, *14*:566, 1961.

Šlot, E.: Spontaneous changes of the stemline cells in human carcinomas. *Neoplasma, 14*:629, 1967.

Spriggs, A. I.: Karyotype changes in human tumor cells. *Br J Radiol, 37*:210, 1964.

Steenis, H., van. Chromosomes and cancer. *Nature* (London), *209*:819, 1966.

Stich, H. F.; Florian, S. F.; Emson, H. E.: The DNA content of tumor cells. I. Polyps and adenocarcinomas of the large intestine of man. *J Nat Cancer Inst, 24*:471, 1960.

Stich, H. F., and Steele, H. D.: DNA content of tumor cells. III. Mosaic composition of sarcomas and carcinomas in man. *J Nat Cancer Inst, 28*:1207, 1962.

Tabata, T.: Karyological studies of human tumors. *Wakayama Igaku, 9*:963, 1959.

Tonomura, A.: Cytological studies of tumors. XXXII. Chromosome analysis in stomach and uterine carcinoma. *J Fac Sci Hokkaido Univ,* Ser VI. 2001, *14*:149, 1959.

Tonomura, A.: The cytological effect of chemicals on tumors. VIII. Observations on chromosomes in a gastric carcinoma treated with carzinophilin. *Gann, 51*:47, 1960.

White, L., and Cox, D.: Chromosome changes in a rhabdomyosarcoma during recurrence and in cell culture. *Br J Cancer, 21*:684, 1967.

Yamada, K.; Takagi, N.; Sandberg, A. A.: Chromosomes and causation of human cancer and leukemia. II. Karyotypes of human solid tumors. *Cancer, 19*:1879, 1966.

Chapter XV

RESPIRATORY TRACT

THE TUMOR of the respiratory tract most frequently analyzed has been carcinoma of the lung. This neoplasm is principally (about 85%) found in males. There seems to be no significant racial predilection. About 80 percent occur in the 40 to 70 years age range. Cancer of the lung has the fastest growing incidence rate of all tumors and consequently environmental pollutants as possible causative factors have become the object of considerable concern.

There are no reports on chromosomes of precanceroses in the respiratory system. Most chromosomal analyses of lung cancer have been performed on pleural effusion cells. Thus, information has been principally obtained from metastatic cells of well-advanced malignancies.

The first report, published by Levan (1956) was concerned with the behavior of a ring chromosome marker in the pleural effusion of an anaplastic adenocarcinoma of the lung. The marker persisted in 100 percent of cells through two samples drawn two weeks apart. The author analyzed variations in structure and transformation of the ring during mitosis and endomitosis.

Further analysis of the above tumor (Levan, 1956) was performed by Ising and Levan (1957). The modal number was 75 and five idiograms showed ring, long acrocentric, and metacentric markers. Chromosome numbers exceeded 140 in 10 percent of cells.

Ishihara et al. (1963) made three samplings of pleural effusion of lung carcinoma. The first sample contained cells with chromosome numbers ranging from 94 to 106. The second sample, obtained six hours after injection of Vincristine, had cells with 72 to 194 chromosomes. The third sample, withdrawn 24 hours following Vincristine application, showed cells with 75 to 129 chromosomes. In the last specimen, two or three acrocentric markers appeared in all cells.

Cox et al. (1965) analyzed the pleural effusion of a pleomorphic anaplastic carcinoma of the lung in a 58-year-old man. The mode was 75. In addition to the set of apparently morphologically undamaged chromosomes, almost all dividing cells contained a number of minute chromatin particles the size of which was near the limit of visibility. These minute chromatin bodies were also found in five tumors from various other sites. The authors suggested their association with micronuclear formation.

Findings of other authors are briefly summarized in Table XXXIV.

Solid primary carcinoma of the larynx was analyzed by Yamada et al. (1966). The modes were 40 (22% cells) and 78 to 80. Karyotypes of modal cells contained haploid numbers in groups D, F, and G. Three long and one minute marker were present in almost all cells examined.

Choudhuri and Choudhuri (1967) analyzed chromosomes from blood lymphocytes of ten patients with neoplasms of the tonsils, pharynx, and epiglottis. In one patient with carcinoma of the tonsil, attenuated secondary constrictions, chromatid breaks, and fragments were observed. The authors did not indicate the number or percent of affected cells.

Fallor et al. (1969) examined bronchial epithelium obtained by bronchoscopic biopsy in one *bronchial adenoma* and four bronchogenic carcinomas. Three metaphases were counted in bronchial adenoma. All had 45 chromosomes variably distributed among the Denver groups. Bronchogenic carcinomas yielded abnormal karyotypes in all cases. The following numbers were observed: the first tumor, 62 to 131; the second tumor, a sharp mode of 60 chromosomes; the third tumor had a mode of 65 chromosomes. Markers were observed in all three tumors with consistency. The fourth carcinoma had abnormal chromosomes, but of too poor quality to construct karyotypes or allow counting.

Several authors have examined the sex chromatin incidence in cells of lung carcinoma in an attempt to correlate its occurrence with the phenotypic sex of the patient or with the sex chromatin incidence of tumors of other sites.

Hanschke and Hoffmeister (1960) examined five bronchogenic carcinomas in males for sex chromatin. An average of 17.4 percent of

TABLE XXXIV
CHROMOSOMES IN RESPIRATORY TRACT TUMORS

Authors	Stemlines	Markers	Site	Remarks
Makino et al. (1959)	—	—	Pulmonary squamous cell carcinoma effusions	near diploid
	46-48 (40%)	—		
	—	—		near diploid
Ishihara et al. (1961)	62, 63	large marker	Pulmonary squamous cell carcinoma effusions	extra A, E, F, G, C
	46 pseudo	large marker		
	80	large marker		extra A, B, D, E, F, G
Spriggs et al. (1962)	30-35	—	Brain metastasis from lung ca.	
	—	—	Solid oat-cell ca. lung	
	80	—	Effusion, oat-cell ca. lung	
	75	—	Effusion, ca. lung	
Ishihara and Sandberg (1963)	46	—	Effusion, ca. lung	
Atkin et al. (1966)	69	—	Ca. bronchus lymphnode metastasis	
	72-75	—	Ca. bronchus lymphnode metastasis	

Authors	Stemlines	Markers	Site	Remarks
Davidson and Bulkin (1966)	—	long acrocentric	Effusion ca. bronchus	extra F, G, C
Yamada et al. (1966)	68 (12%)		?	
Miles (1967)	57-62	large meta- and submetacentric	solid ca. lung	
Sandberg et al. (1967)	69	long acrocentric and minute chr.	effusion ca. lung	extra A, C, E missing D, G
Kotler and Lubs (1967)	72	large sub-metacentric	ca. lung	extra C, E, F, and G
Jackson (1967)	66	—	effusion, ca. lung	
	40	—	effusion, ca. lung	
	114	—	effusion, ca. lung	
	84	—	effusion, ca. lung	
	76	—	effusion, ca. lung	

cells was noted to be chromatin-positive. Comparing this data with samples from 50 carcinomas of the uterine cervix, in which an average of 67.2 percent of cells were chromatin positive, the author concluded that cancer in general preserves the cytomorphologic sex characteristics of the host tissue. Hanschke and Hoffmeister (1960), examining 20 bronchogenic carcinomas in women, confirmed the findings noted above. Positive sex chromatin was observed in 64.5 percent of the cells examined.

Similar results were obtained by Baradnay et al. (1968) in 37 carcinomas of the lung in female patients. In 20 cases, 90 to 100 percent of tumor cells contained Barr bodies. In no case did less than 30 percent of the cells have a clearly visible sex chromatin body.

The chromosome number in lung and larynx carcinomas varied greatly as with malignant neoplasms in general. Modal numbers were usually near diploid and triploid values and ranged from 40 to 114. Stemlines detected in two solid tumors (68; 78 to 80) did not differ from most stem cell numbers of effusion metastatic cells. In the karyotypes of malignant cells, distribution of chromosomes (according to the Denver grouping) did not reveal any consistent pattern. However, groups E, F, and C most often contained supernumerary chromosomes and the D group frequently lacked several chromosomes.

REFERENCES

Atkin, N. B.; Mattinson, G.; Baker, M. C.: A comparison of the DNA content and chromosome number of fifty human tumors. *Br J Cancer, 20*:87; 1966.

Baradnay, G.; Monus, Z.; Kulka, F.: Sexchromatin-Untersuchungen in weiblichen Lungenkrebs-Fällen. *Zbl Allg Path Anat, 111*:275, 1968.

Choudhuri, R. D., and Choudhuri, A.: Chromosome findings in leucocytes of patients with cancer of the throat region. *Indian J Cancer, 4*:180, 1967.

Cox, D.; Yuncken, C., Spriggs, A. I.: Minute chromatin bodies in malignant tumours of childhood. *Lancet, 2*:55, 1965.

Davidson, E., and Bulkin, W.: Long marker chromosome in bronchogenic carcinoma. *Lancet, 2*:227, 1966.

Fallor, W. H.; Gordon, M.; Maczala, O. A.: Chromosomes in bronchoscopic biopsies. *Cancer, 24*:198, 1969.

Haemmerli, G.; Fjelde, A.; Zweidler, A.; Sträuli, P.: Heterologous transplantation, chromosome analyses and DNA measurements of the human carcinoma tissue culture line, H. Ep. #2. *J Nat Cancer Inst, 36*:673, 1966.

Hanschke, H. J., and Hoffmeister, H.: Die zellkernmorphologische Geschlechtsbestimmung beim Bronchialcarzinom der Frau. Zbl Allg Path Anat, 101:99, 1960.
Ishihara, T.; Moore, G. E.; Sandberg, A. A.: Chromosome constitution of cells in effusions of cancer patients. J Nat Cancer Inst, 27:893, 1961.
Ishihara, T.; Kikuchi, Y.; Sandberg, A. A.: Chromosomes of twenty cancer effusions—Correlation of karyotypic, clinical and pathologic aspects. J Nat Cancer Inst, 30:1303, 1963.
Ishihara, T., and Sandberg, A. A.: Chromosome constitution of diploid and pseudodiploid cells in effusions of cancer patients. Cancer, 16:885, 1963.
Ising, U., and Levan, A.: The chromosomes of two highly malignant human tumors. Acta Path Microbiol Scand, 40:13, 1957.
Jackson, J. F.: Chromosome analysis of cells in effusions from cancer patients. Cancer, 20:537, 1967.
Kotler, S., and Lubs, H. A.: Comparison of direct and short-term tissue culture technics in determining solid tumor karyotypes. Cancer Res, 27:1861, 1967.
Levan, A.: Self-perpetuating ring chromosomes in two human tumours. Hereditas, 42:366, 1956.
Makino, S.; Ishihara, T.; Tonomura, A.: Cytological studies of tumors. XXVII. The chromosomes of thirty human tumors. Z Krebsforsch, 63:184, 1959.
Miles, C. P.: Chromosome analysis of solid tumors. II. Twenty-six epithelial tumors. Cancer, 20:1274, 1967.
Sandberg, A. A.; Yamada, K.; Kikuchi, Y.; Takagi, N.: Chromosomes and causation of human cancer and leukemia. III. Karyotypes of cancerous effusions. Cancer, 20:1099, 1967.
Spriggs, A. I.; Boddington, M. M.; Clarke, C. M.: Chromosomes of human cancer cells. Br Med J, 2:1431, 1962.
Stich, H. F., and Steele, H. D.: DNA content of tumor cells. III. Mosaic composition of sarcomas and carcinomas in man. J Nat Cancer Inst, 28:1207, 1962.
Yamada, K.; Takagi, N.; Sandberg, A. A.: Chromosomes and causation of human cancer and leukemia. II. Karyotypes of human solid tumors. Cancer, 19:1879, 1966.

Chapter XVI

NERVOUS SYSTEM

THE TUMORS TO BE CONSIDERED here are of primary neurogenic origin as well as those allied to the nervous system. Their nomenclature and classification are probably the most difficult of all tumors. In several cases presented, the classification is somewhat arbitrary since some authors have failed to elaborate on the histology or location of the tumor analyzed. The neoplasms will be arranged in histopathologic groups based on the simplified classification of Minckler (1961):

    I. Chordoma
   II. Primary intramedullary tumors
      A. Gliomas
      B. Nerve cell tumors
  III. Primary extramedullary neurogenic tumors
      A. Supporting cell tumors
      B. Tumors of peripheral nerve cells and their processes
      C. Miscellaneous peripheral neurogenic tumors
  IV. Meningiomas
   V. Tumors of cerebral vascular and perivascular structures
      A. Tumors of blood vessels
      B. Tumors of neurodermal supporting tissue
  VI. Mixed tumors
 VII. Hypophyseal tumors
VIII. Metastatic tumors

The groups in which no anomaly has been detected such as chordomas will not be considered.

PRIMARY INTRAMEDULLARY NEUROGENIC TUMORS

Gliomas (Supporting Cells Tumors)

Chromosome analyses have shown a rather narrow range of chro-

TABLE XXXV
PRIMARY INTRAMEDULLARY NEUROGENIC TUMORS
A. GLIOMAS (SUPPORTING CELL TUMORS)

Author	Mode	Range of Counts	Markers	Specimen	Diagnosis
Spriggs et al. (1962)	80	—	—	Primary	Glioma of brain
Erkman and Conen (1964)	46 pseudo	—	—	Primary	Ependymoma
	—	109-113	—	Primary	Astrocytoma III
	46 and 46 pseudo	tetraploidy	—	Primary	Astrocytoma III
	46	62-67	—	Primary	Astrocytoma IV
	47	46-80	—	Primary	Astrocytoma IV
	47	—	—	Primary	Astrocytoma IV
	46	—	—	Primary	Astrocytoma IV
		only 46 norm.		Primary	Medulloblastoma
Lubs and Salmon (1965)	49	45-51	Double minutes in 100%	Primary	Medulloblastoma
	92	—	—	Primary	Oligodendroglioma
	42-47				
	86-89	—	Acrocentric and acentric in both stemlines	Primary	Glioblastoma multiforme
Cox et al. (1965)	—	78-85	Double minutes in 100%	Primary	Medulloblastoma
Lubs et al. (1966)	49	45-51	Double minutes in 100% metacentric	Primary	Medulloblastoma

TABLE XXXV—(Con't.)
PRIMARY INTRAMEDULLARY NEUROGENIC TUMORS
A. GLIOMAS (SUPPORTING CELL TUMORS)

Author	Mode	Range of Counts	Markers	Specimen	Diagnosis
Hansteen (1967)	48	—	Double minutes in one cell	Primary	Astrocytoma
	46 pseudo	45-92		Primary	Astrocytoma
	47	Few polyploid		Primary	Astrocytoma
	46 pseudo	—	Resembling #2	Primary	Astrocytoma
	49-50	—	Small metacentric	Primary	Astrocytoma
	—	43-46		Primary	Astrocytoma
Conen and Falk (1967)	46	—	—	Cultured	Medulloblastoma
	46	—	—	Cultured	Medulloblastoma
	46	—	—	Cultured	Astrocytoma
	46	—	—	Cultured	Ependymoma
	46	—	—	Cultured	Ependymoma
	46	—	—	Cultured	Papilloma chorioideum
	46	—	—	Cultured	Glioma of optic nerve
Kotler and Lubs (1967)	51	—	Long marker	Primary and Culture	Medulloblastoma
		Wide variation	—	Primary and Culture	Glioblastoma
Kucheria (1968)	46 pseudo	—	Double minutes in 33%	Cultured (5 weeks)	Subependymal glioma
Cox (1968)	46	45-46	—	Primary	Subependymal glioma
	46	37-46	—	Primary	Subependymal glioma
	47	—	Large acrocentric	Primary	Subependymal glioma
	46 pseudo	—	—	Primary	Medulloblastoma
	84 and 86	53-97	1 or 2 small fragments	Primary	Astrocytoma

TABLE XXXVI
PRIMARY INTRAMEDULLARY NEUROGENIC TUMORS
B. NERVE CELL TUMORS

Author	Mode	Range of Counts	Markers	Specimen	Diagnosis
Spriggs et al. (1962)	46	46-92	—	Pleural effusion	Neuroblastoma
Gagnon et al. (1962)	48	—	2 large markers	Culture	Neuroblastoma
Lele et al. (1963)	46	—	All cells 15q-	Primary	Retinoblastoma
	46	—			
	46	—			
	46	—	Normal	Blood	Retinoblastoma
	46	—			
	46	—			
	46	—			
Wiener et al. (1963)	12 patients with hereditary form				
	5 patients with sporadic tumor				
	All normal 46			Blood	Retinoblastoma
Day et al. (1963)	48	Double Trisomy: XXX and G₁ trisomy		Blood	Retinoblastoma in Down's syndrome
Makino et al. (1965)	46		2 cells with long submetacentric	Pleural effusion	Neuroblastoma
Brewster and Garrett (1965)	48			Bone marrow	Neuroblastoma
Cox et al. (1965)	46 pseudo	37-58	Double minutes in 80%	Primary	Neuroblastoma
	54	49-55	Small and large acrocentric	Primary	Neuroblastoma
			Large submetacentric		
			Double minutes in 50%		
	75	56-78	Double minutes in majority	Primary	Neuroblastoma
Conen and Falk (1967)	46		Six tumors, all normal karyotypes	Culture	Neuroblastoma
Miles (1967)	46			Primary	Neuroblastoma
Levan et al. (1968)		Around diploid	No large long arms Double minutes in majority	Primary	Neuroblastoma
Cox (1968)	53	40-54	Large dicentric Large submetacentric Large acrocentric	Primary	Neuroblastoma

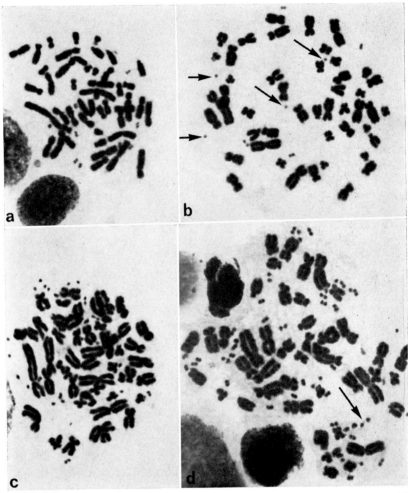

Figure 21. Four metaphases from a male patient with neuroblastoma. All contain 46 normal chromosomes and varying numbers of minute chromosomes: (a) 9; (b) 6; (c) 18; (d) 45. (Courtesy of Dr. A. Levan et al., *J Nat Cancer Inst*, 41:1377, 1968.

mosome counts, well-defined stemlines, and not many prominent markers (see Table XXXV). In no case was extreme polyploidy observed nor extensive disorganization of karyotype. Modal numbers in the majority of cases were in the range of diploidy. Remarkable findings were double minute chromosomes in metaphases of five primary tumors (Lubs and Salmon, 1965; Cox et al., 1965; Lubs et al., 1966; and Kucheria, 1968).

Nerve Cell Tumors

Noteworthy is the presence of double minute chromosomes. They were observed in four of six primary solid neuroblastomas by Cox et al. (1965) and Levan et al. (1968) (See Fig. 21). A number of marker chromosomes were described but none was seen consistently in more than one of each tumor investigated (see Table XXXVI).

Twenty-four patients having retinoblastoma were examined, in most cases the chromosomes being obtained from peripheral blood cultures. All except two had a normal complement of 46 chromosomes.

In the case of Day et al. (1963), the patient had double trisomy: XXX and trisomy 21. This female had typical phenotypic signs of Down's syndrome, no breast development, but normal external genitals. She had retinoblastoma of the right eye. A second abnormal finding was noted in the case of Lele et al. (1963). The patient had retinoblastoma with deletion of one-third of the long arm of AD chromosome in all cells of the primary tumor, dermal tissue, conjunctiva, and blood lymphocytes. Tumor tissue had an additional number of hypodiploid cells. Parents and siblings did not exhibit any chromosomal abnormality. Quite different patterns emerged when *probands were selected for deletion of long arm of D chromosome (Dq-syndrome)*. Including the case of Lele et al., 12 patients with Dq- were examined and, surprisingly, seven of them had retinoblastoma; five had bilateral, one had unilateral retinoblastoma, and in one this information was not available (Taylor, 1970, Grave et al., 1971). This striking finding led Taylor to suggest the hypothesis that the retinoblastoma locus is situated in the proximal part of the long arm of one of the D chromosomes.

Unfortunately, reliable identification of the particular deleted chromosome responsible for the retinoblastoma was not possible. It may be assumed that in five of the 12 cases of Dq- not exhibiting retinoblastoma, a different D chromosome was involved than the one carrying the retinoblastoma locus(?).

Retinoblastoma occupies an unusual position among malignant tumors being inherited in many cases as an autosomal dominant trait. Generally, the data available suggest two types of retinoblastoma: (a) familial and (b) sporadic. Familial cases most often are bilateral but not in all cases (Smith and Sorsby, 1958). The penetrance of the gene is not complete and skipping of generations is common (Macklin, 1959). If a single gene mechanism is involved in the etiology of this tumor, more data on chromosomes of tumor tissue in early stages of malignant transformation will be of special interest.

PRIMARY EXTRAMEDULLARY NEUROGENIC TUMORS
Supporting Cell Tumors

This group of neurogenic tumors is represented by a single case of neurofibromatosis with abnormal cytogenetic findings (Milcu et al., 1961). Of 35 cells analyzed, 14 were diploid and the rest polyploid. Chromosome preparation was carried out on bone marrow cells.

Miscellaneous Peripheral Neurogenic Tumors

Eight meningiomas were analyzed (Sang and Singer, 1967). Specimens were cultured five to nine days. In all eight tumors, a G-chromosome was missing and no structural abnormality was observed! None of the tumors had a normal 46 stemline: modal numbers were as follows: 45, 45, 45, 44, 46 pseudodiploid, 42, 42. In one tumor with mode 42 and missing D, C, E, and G, chromosomes, 30 of 48 cells contained double minutes.

Erkman and Conen (1964) examined a solid, primary cerebellar sarcoma with mode 49, cerebellar metastases from a rhabdomyosarcoma with counts ranging from 47 to 88, cerebellar metastases from a carcinoma with counts 44 to 64 and a meningeal sarcoma with a pseudodiploid 46 mode.

Yamada et al. (1966), reported primary "carcinoma of the brain." Chromosome counts ranged from 38 to 48 with a 46 pseudodiploid mode. Modal cells had a missing C_6, supernumerary C_{10} (?) and no marker.

In reviewing chromosomal findings of tumors of the nervous system, it would appear that there is no distinct marker which may be regarded as a common or characteristic feature. *Minute chromosomes* were found in this group of tumors much more frequently than in tumors of other sites. They are present in the normal karyotype of several species of animals and plants. In human tumors, their origin and meaning are still obscure. They occur in cells without other evidence of chromosome breakage which could be caused by radiation, viruses, or by drugs. Their size varies, the largest mass approaches a G-chromosome in size, but usually they are much smaller, being spherical, and about 0.5μ in diameter. They are usually capable of replication and their persistence suggests that they possess a centromere or a portion thereof. White and Cox (1967) maintained rhabdomyosarcoma cells with double minutes in culture for eight months.

They were also observed in acute monomyelocytic leukemia (see Chap. VIII).

In some instances it has been suggested that minutes might be associated with micronuclear formation (Cox *et al.*, 1965; Kucheria, 1968). Lubs *et al.* (1966) considered the possibility of unequal division of a ring chromosome or misdivision of the centromere of a 17 to 18 chromosome which occurred near the short arm end of the centromere.

Levan *et al.* (1968) performed detailed analysis of 109 neuroblastoma cells, 104 of which had minute chromosomes. They observed that the mitotic mechanism of minutes was much less stable than that in other chromosomes suggesting that their functional centromere is less efficient than the centromere of ordinary chromosomes. They are probably genetically empty or weakly active. The authors suggested that minutes most likely originate from simultaneous breaks of several centromeric regions. Many centric fragments may acquire neocentric activity.

An important observation is that in some higher plants certain environmental factors have been shown to influence either elimination or accumulation of minutes. Since double minute chromosomes are distinctively more frequent in neurogenic tumors than in all other cancers, one may speculate that this could be caused by unknown specific characteristics of neurogenic tumors which would favor accumulation and persistence of minutes.

It remains moot whether the appearance of double minutes is associated with such specific features of neurogenic tumors as for instance striking radiosensitivity of some neoplasms (medulloblastoma) or the rare ability to regress and eventually disappear spontaneously (neuroblastoma).

REFERENCES

Brewster, D. J., and Garrett, J. V.: Chromosome abnormalities in neuroblastoma. *J Clin Path, 18*:167, 1965.

Conen, P. E., and Falk, R. E.: Chromosomes studies on cultured tumors of nervous tissue origin. *Acta Cytol, 11*:86, 1967.

Cox, D.; Yuncken, C.; Spriggs, A. I.: Minute chromatin bodies in malignant tumours of childhood. *Lancet, 2*:55, 1965.

Cox, D.: Chromosome studies in 12 solid tumours from children. *Br J Cancer, 22*:402, 1968.

Day, R. W.; Wright, S. W.; Koons, A.; Quigley, M.: XXX/ 21-trisomy and retinoblastoma. *Lancet, 2*:154, 1963.

Erkman, B., and Conen, P. E.: Consistent pseudodiploid and near diploid karyotypes in three intracranial tumors. *Am J Pathol, 44*:18a, 1964 (abstract).

Gagnon, J.; Dupal, M. F.; Katyk-Longtin, N.: Anomalies chromosomiques dans une observation de sympathome congénital. *Rev Canad Biol, 21*:145, 1962.

Grace, E.; Drennan, J.; Colver, D.; Jordan, R. R.: The 13q-deletion syndrome. *J Med Genet, 8*:351, 1971.

Hansteen, I.: Chromosome studies in glial tumours. *Europ J Cancer, 3*:183, 1967.

Kotler, S., and Lubs, H. A.: Comparison of direct and short-term tissue culture technics in determining solid tumor karyotypes. *Cancer Res, 27*:1861, 1967.

Kucheria, K.: Double minute chromatin bodies in a subependymal glioma. *Br J Cancer, 22*:696, 1968.

Lele, K. P.; Penrose, L. S.; Stallard, H. B.: Chromosome deletion in a case of retinoblastoma. *Ann Hum Genet, 27*:171, 1963.

Levan, A.; Manolov, G.; Clifford, J.: Chromosomes of human neuroblastoma: a new case with accessory minute chromosomes. *J Nat Cancer Inst, 41*:1377, 1968.

Lubs, H. A., Jr., and Salmon, J. H.: The chromosomal complement of human solid tumors. II. Karyotypes of glial tumors. *J Neurosurg, 22*:160, 1965.

Lubs, H. A., Jr.; Salmon, J. H.; Flanigan, S.: Studies of a glial tumor with multiple minute chromosomes. *Cancer, 19*:591, 1966.

Macklin, M. T.: Inheritance of retinoblastoma in Ohio. *Arch Ophthal, 62*:842, 1959.

Makino, S.; Safini, T.; Mitani, M.: Cytological studies on tumors. XLIII: A chromosome condition of effusion cells from a patient with neuroblastoma. *Gann, 56*:127, 1965.

Milcu, St. M.; Stanescu, V.; Ionescu, V.; Maximilian, C.: Mozaic cromozomial: 46 cromozomi si poliploidie—in neurofibromatozo. *Stud Cercet Endocr, 12*:745, 1961.

Miles, C. P.: Chromosome analysis of solid tumors. *Cancer, 20*:1253, 1967.

Minckler, J.: Nervous system. In Anderson, W. A. D. (Ed.): *Pathology*, 4th ed. St. Louis, C. V. Mosby, 1961, p. 1311.

Smith, S. M., and Sorsby, A.: Retinoblastoma. Some genetic aspects. *Ann Hum Genet, 23*:50, 1958.

Spriggs, A. I.; Boddington, M. M.; Clarke, C. M.: Chromosomes of human cancer cells. *Br Med J, 2*:1431, 1962.

Taylor, A. I.: Dq-, Dr, and retinoblastoma. *Humangenetik, 10*:209, 1970.

White, L., and Cox, D.: Chromosome changes in a rhabdomyosarcoma during recurrence and in cell culture. *Br J Cancer, 21*:684, 1967.

Wiener, S.; Reese, A. B.; Hyman, G. A.: Chromosome studies in retinoblastoma. *Arch Ophthal, 26*:311, 1963.

Yamada, K.; Takagi, N.; Sandberg, A. A.: Chromosomes and causation of human cancer and leukemia. II. Karyotypes of human solid tumors. *Cancer, 19*:1879, 1966.

Zang, K. D., and Singer, H.: Chromosomal constitution of meningiomas. *Nature, 216*:84, 1967.

Chapter XVII

URINARY SYSTEM

MOST NEOPLASMS of the urinary system which have been karyotyped have been carcinomas of the urinary bladder. Tumors primary in this location have several characteristic features. They usually occur between 50 and 70 years of age, are nearly three times more frequent in males, and generally exhibit relatively slow progressive growth. Frequently they have their origin in papillomas of the bladder.

Not uncommonly the normal human bladder epithelial cell may exhibit polyploidy. Although most cells are diploid, tetraploid forms occur with high frequency and even octoploid cells are occasionally found. This type of physiological polyploidy has also been noted in the urinary bladder of other mammals. In contrast, the epithelium of normal human ureter and urethra is typically diploid. The transitional cell epithelium of the bladder has a markedly low mitotic activity in the resting stage. Pringle and Williams (1967) ascertained only two mitotic figures among 16,679 epithelial cells examined. Confirming data were also collected by Cooper et al. (1969) who used radioactive labeling of DNA to observe its replication.

Polyploid cells of bladder epithelium are self-maintaining dividing cells, probably located more superficially than the smaller, deeper positioned, diploid cells.

From the following data on chromosomal changes and DNA content of bladder carcinoma, one may deduce that the tumor most likely originates from these small diploid cells rather than from the polyploid superficial cells.

The first report on the chromosomes of bladder carcinoma was published by Spriggs et al. (1962). The authors examined peritoneal effusions from two patients. In the first case there was a mode of about 47 to 49 chromosomes. The second case was characterized by scattered values without a clear stemline.

Yamada et al. (1966) analyzed two cases of solid primary bladder carcinoma. Ninety-five cells from the first case had chromosome numbers ranging between 48 and 87 and a clear mode of 82. All cells examined contained a ring marker. Extra chromosomes were persistent in all groups except the B group. The second tumor had a mode of 45 (44% cells), chromosome numbers ranging between 40 and 49.

Tavares et al. (1966) measured optical density of Feulgen-stained nuclei of 62 bladder carcinomas. Thirty-five tumors were diploid and tetraploid. Thirteen (37%) of the patients died between one month and nine years following surgery, the average survival time being 33.6 months. Twenty-seven tumors with triploid or hexaploid constitution had markedly worse prognosis. Twenty-one patients died between one month and six years after surgery, which represents an average survival time of 18.1 months.

Comparing the above data with the mean life expectancy of a random Portuguese population, the authors found a median reduction of life expectancy in diploid and tetraploid bladder tumors to be 55 percent in contrast to 88 percent in those with triploid and hexaploid tumors.

Lamb (1967) analyzed chromosomes from 30 primary carcinomas of the bladder. Well-differentiated tumors were diploid and a high proportion (54%) of examined cells formed the stemline.

Moderately well-differentiated tumors showed a frequently near-tetraploid mode. Tumors with near-diploid cells in this group showed no mucosal invasion, in contrast to four invasive tumors all having near-tetraploid counts.

Poorly differentiated carcinomas were mainly hypotetraploid, having a wider range of counts than well- and moderately well-differentiated tumors. Some cells were markedly hypodiploid and others near octoploid. Though detailed karyotyping was not done, large and minute markers were frequently observed in most specimens, except in those from well-differentiated tumors.

The proportion of stem cells was lower in poorly differentiated tumors (17%) than in well-differentiated examples (54%) indirectly suggesting an association between modal number and tumor inva-

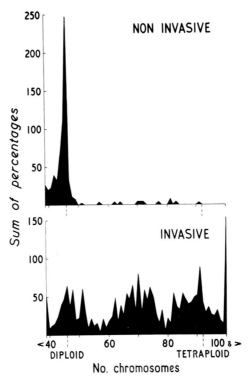

Figure 22. Distribution of chromosome numbers of cells from eight cases of noninvasive and 22 cases of invasive urinary bladder tumors. Expressed is the sum of the percentage of cells counted for each case. (Courtesy of Dr. D. Lamb, *Br Med J*, 1:273, 1967.)

siveness. All eight noninvasive tumors had near-diploid modes in contrast to 22 invasive tumors (Fig. 22).

Cox (1968) examined rhabdomyosarcoma of the bladder in a three-month-old girl. The mode was 46 (70%) and of 12 karyotypes 9 showed normal diploid constitution.

Similar results were obtained by Cooper *et al.* (1969) by measuring the DNA content of 60 bladder tumors. They found that as transitional cell carcinoma becomes more invasive, there is usually progressive shift of the modal DNA content toward higher values. Chromosomal analysis showed that even well-differentiated tumors had chromosomal abnormalities. The authors injected colcemid (10 mg/I.V.) three hours before resection.

The above results were confirmed by Levi et al. (1969). The authors measured DNA content and in vitro incorporation of tritiated thymidine in 31 transitional cell carcinomas of the bladder. Tumors were classified into three groups: (a) differentiated papillary tumors of low malignancy, (b) invasive papillary tumors, and (c) solid tumors. In general, progression of ploidy from diploid in low malignancy tumors to a wide range of polyploid values in anaplastic tumors was observed.

Atkin and Baker (1969) found a mode of 45 and missing chromosomes in groups B, D, and G in carcinoma of the bladder.

Few other tumors of the urinary tract have been examined chromosomally: nephroblastoma and carcinoma of the prostate, urethra, and ureter.

Cox (1966) examined seven cases of untreated primary nephroblastoma. All specimens were obtained from children and their analysis showed low diploid, pseudodiploid, or near-diploid counts with surprising consistency. Neither highly polyploid cells nor abnormal markers were found. Counts ranged from 44 in case 1 to 48 in case 2. Of eight karyotypes, two were normal diploid and the rest contained extra chromosomes in groups A and C. Case 3 had a mode of 46. Twenty-two karyotypes were constructed, all being normal except for three cells with 45 chromosomes. Each of these hypodiploid cells was obtained from a different site within the tumor. All cells counted as well as all 13 karyotypes showed a normal set of 46 chromosomes. Case 5 exhibited 46 chromosomes in all eight cells counted and three karyotypes had a normal diploid set. In case 6, all 15 counted cells had a chromosome number ranging from 55 to 58. Five karyotypes showed a missing G and extra A and C chromosomes. Case 7 was clearly pseudodiploid. All 14 karyotypes had 46 chromosomes with a missing A_1 and an extra chromosome in the C group. All 44 cells contained 46 chromosomes.

Histological examination of cells from case 4 confirmed that cells with normal diploid karyotype were tumor cells and not host cells.

Tavares et al. (1966) determined the DNA content of prostatic carcinomas. Diploid and tetraploid values were found in 26 cases. Triploid and hexaploid values were determined in nine cases. The mean survival period for the diploid and tetraploid group was 7.4 years, i.e. significantly higher than 4.1 years in the triploid and hexaploid group.

An interesting finding was the difference in response to estrogens in the two groups: All but one of the nine triploid and hexaploid tumors did not respond to estrogen therapy, sharply contrasting with 22 of 24 diploid and tetraploid tumors which responded well.

Levi (1969) examined the DNA content in five carcinomas of the ureter and one carcinoma of the urethra. All tumors were in diploid range except for one highly invasive tumor of hypo-octoploid composition.

Despite the fact that some of the above reports lack detailed histologic diagnosis, chromosomal analysis in carcinoma of the urinary bladder seems to yield some practical information. Attempts to elaborate meaningful prognostic criteria suggest an association between high ploidy and invasiveness, eventually between ploidy of the tumor and prognosis concerning its recurrence. Significantly, analogous conclusions were reached independently by several authors, namely Shigematsu (1965), Tavares et al. (1966), Lamb (1967), Levi et al. (1969) and Cooper et al. (1969) on the basis of analysis of more than 180 tumors in different stages of malignancy.

REFERENCES

Atkin, N. B., and Baker, M. C.: Possible differences between the karyotypes of preinvasive lesions and malignant tumors. Br J Cancer, 23:329, 1969.

Cox, D.: Chromosome constitution of nephroblastomas. Cancer, 19:1217, 1966.

Cox, D.: Chromosome studies in 12 solid tumours from children. Br J Cancer, 22:402, 1968.

Cooper, E. H.; Levi, P. E.; Anderson, C. K.; Williams, R. E.: The evolution of tumour cell populations in human bladder cancer. Br J Urol, 41:714, 1969.

Lamb, D.: Correlation of chromosome counts with histological appearances and prognosis of transitional-cell carcinoma of bladder. Br Med J, 1:273, 1967.

Levi, P. E.; Cooper, E. H.; Anderson, C. K.; Williams, R. E.: Analyses of DNA content, nuclear size and cell proliferation of transitional cell carcinoma in man. Cancer, 23:1074, 1969.

Makino, S.; Ishihara, T.; Tonomura, A.: Cytological studies of tumors, XXVII. The chromosomes of thirty human tumors. Z Krebsforsch, 63:184, 1959.

Pyrah, L. N.; Raper, F. P.; Thomas, G. M.: Report of a follow-up of papillary tumors of the bladder. *Br J Urol, 36*:14, 1964.

Pringle, J. A. S., and Williams, R. F.: Mitoses in human bladder biopsies. *Annual Rep Brit Empire Cancer Campaign, 11*:237, 1967.

Spriggs, A. I.; Boddington, M. M.; Clarke, C. M.: Chromosomes of human cancer cells. *Br Med J, 2*:1431, 1962.

Shigematsu, S.: Significance of the chromosome in vesicle cancer. In *Proc XIII Cong Soc Int d'Urologie,* Vol. 2, 1965, p. 111.

Tavares, A. S.; Costa, J.; Carvalho de, A.; Reis, M.: Tumor ploidy and prognosis in carcinomas of the bladder and prostate. *Br J Cancer, 20*:438, 1966.

Yamada, K.; Takagi, N.; Sandberg, A. A.: Chromosomes and causation of human cancer and leukemia; II. Karyotypes of human solid tumors. *Cancer, 19*:1879, 1966.

Chapter XVIII

MISCELLANEOUS TUMORS

IN THE PREVIOUS chapters enough examples have been presented to provide the reader with sufficient understanding of certain rules concerning the cytogenetics of tumors, such as evolution of stemlines, nonspecificity of most chromosomal changes, similarity of karyotypes in malignancies of different sites, and dissimilarity in karyotypes of tumors of identical sites or histologic picture.

There would appear to be little reason to present detailed review of all reports on tumors of miscellaneous histology and sites not included in previous chapters. As difficult as they are in histologic nature, so is their marked variability in karyotypes.

As specific examples we should like to discuss *melanoma* and *malignant thyroid* tumors since these have been reported in larger series. For example, 20 cases of *malignant melanoma* were studied The majority of specimens examined were metastatic cells either in effusion or from bone marrow or lymph nodes. A variety of markers was observed. The chromosome number of stemline cells varied extensively and no unusual characteristics were observed with consistency.

Forteza Bover and Baguena Candela (1966) examined cells obtained from metastasis of melanoma to lymph nodes. Preparation was direct, i.e. without culturing. There was no karyotypic consistency and no apparent predominant cell line. A giant submetacentric marker, ring chromosome, and medium-size telocentric marker were found in several cells. There were ten chromosomes resembling those of the F group in some cells and two missing chromosomes from the G-Y group in others.

The largest series was reported by Miles (1967) who analyzed chromosomes in ten melanomas, nine metastatic and one which was primary. Two of the specimens were characterized by prominent stemlines: 42 chromosomes and 78 chromosomes, respectively. Twelve

TABLE XXXVII
CHROMOSOMES OF MELANOMA

Author	Mode	Range of Counts	Markers	Specimen
Hsu (1954)	Around 46	40-100	—	Peritoneal effusion
Ishihara et al. (1961)	46 pseudo	41-49	2 large metacentric 1 very small	Pleural effusion
	44	43-48	—	Peritoneal effusion
Spriggs et al. (1962)	42	37-80	—	Pleural effusion
Sandberg et al. (1967)	59 and 62	52-248	1 or 2 long acrocentrics	Pleural effusion
Forteza and Baguena (1966)	79-86	79-320	Large submetacentric	Lymphnode
Miles (1967)	42	—	Marker of F size	Solid
	78	—	Ring	Solid
	—	40-92	Several, varied	Solid metastases
	—	50-100	Secondary constrictions	Solid
	55 tumors—all variable in number and karyotpe			Solid
		Variable chromosome numbers		Primary solid
Kong-oo Goh (1968)	46	25% aneuploids in marrow 29% in blood	Acrocentric	Bone marrow and blood cultured
	46	16% aneuploid	Acrocentric	Bone marrow cultured
Atkin and Baker (1969)	66	—	—	Solid
Huang et al. (1969)	46	—	—	Cell line RPMI 1348

of the modal cells with 42 chromosomes were karyotyped and all contained a small condensed F-sized marker. This marker was not seen in seven karyotypes constructed from stemline cells with 78 chromosomes, all of which had a ring chromosome. The remaining eight specimens, including the primary melanoma, contained abnormal numbers of chromosomes, variable markers, and structural rearrangements with no consistent pattern. Review of these and other cases is presented in Table XXXVII.

Thyroid tumors have been investigated by several authors. Miles and Gallagher (1961) cultured metastatic thyroid carcinoma obtaining normal (probably host cell) karyotypes. Garneau (1964) published his data on spectrophotometry of thyroid fibrosis, goiter, adenomas, and oncocytoma, but did not analyze chromosomes. Socolow *et al.* (1964) studied chromosomes of one malignant and five benign tumors of the thyroid gland. Three follicular adenomas were analyzed, 199 cells counted, and 100 cells karyotyped. All three tumors had a clear mode of 46 with normal karyotype. Abnormal chromosome counts were detected in about 20 percent of cells. When these cells were karyotyped, no consistent pattern was seen and the gains and losses involved chromosomes of different groups. Two cases of microfollicular adenoma showed patterns similar to those of follicular adenoma.

One follicular adenocarcinoma differed sharply from benign tumors of the series. Of 100 cells counted, the majority had 52 to 59 chromosomes with gains in the B, C, E, F, and G groups. Of 29 karyotypes, only one was normal diploid; the rest displayed a variable picture with no typical markers, and in 40 percent, one or two centric fragments were present.

The report by Haemmerli (1970) dealt with DNA photometric analysis of 371 nodular lesions of the human thyroid including 17 cases of carcinoma. Irregularly increased DNA values were found in 75 percent of nodules with microfollicular structure, in 81 percent of lesions of oncocytic structure, and in 82 percent of carcinomas. Chromosomal analysis was performed in eleven cases. The largest deviation from diploid number was found in three microfollicular lesions. Thirty-eight percent of the cells in one tumor and 34 percent of the cells of the other were aneuploid; one had a stemline of 60 chromosomes (53%). The third, a solid carcinoma, had a stemline of 46 chromosomes (81%).

These examples of melanomas and thyroid tumors do not permit any deductions. It is needless to say that similar variability has been observed in an inordinate number of case reports on various other human tumors not discussed in the previous chapters. These reports include follicular lymphosarcomas and leiomyosarcoma (Miles, 1967), liposarcoma (Hsu, 1954; Ishihara et al., 1962), rhabdomyosarcoma (Cox et al., 1965; Miles, 1966; White and Cox, 1967; Cox, 1968), osteosarcoma (Stich and Steele, 1962; Ishii, 1965), reticulosarcoma (Makino, 1959; Ishii, 1965; Atkin et al., 1966; Spiers and Baikie, 1967; Miles, 1967), chondrosarcoma and fibrosarcoma (Ishii, 1965), spleen cancer (deGrouchy et al., 1963), sacrococcygeal teratoma (Laurent et al., 1968), and tumors of peritoneum (Ishihara et al., 1963; Vincent et al., 1964; Sandberg et al., 1966, 1967; Cox, 1968).

One may reasonably assume that information of some value could be extracted even from data of such diversity by employing modern methods of chromosomal differentiation such as techniques for revealing different forms of satellite DNA, constitutive heterochromatin. At this point in time, however, their meaning does not exceed the significance of case reports in general.

REFERENCES

Atkin, N. B., and Baker, M. C.: Chromosome abnormalities as primary events in human malignant disease: Evidence from marker chromosomes. *J Nat Cancer Inst, 36*:539, 1960.

Atkin, N. B., and Baker, M. C.: Possible differences between the karyotypes of preinvasive lesions and malignant tumours. *Br J Cancer, 23*:329, 1969.

Cox, D.: Chromosome studies in 12 solid tumours from children. *Br J Cancer, 22*:402, 1968.

Cox, D.; Yuncken, C.; Spriggs, A. I.: Minute chromatin bodies in malignant tumours of childhood. *Lancet, 2*:55, 1965.

Forteza Bover, G., and Baguena Candela, D.: Analisis cromosomico de las cellulas metastaticas de un melanoblastoma maligno obtenidas mediante puncion ganglionar. *Sangre, 11*:161, 1966.

Garneau, R.: Analyses quantitatives cytospectrophotométriques de l'ADN "in situ" dans la thyroide humaine. *Laval Med, 35*:188, 1964.

Grouchy, de J.; Vallée, G.; Lamy, M.: Analyse chromosomique directe de deux tumeurs malignes. *CR Acad Sci* (Paris), *256*:2046, 1963.

Haemmerli, G.: Zytophotometrische und zytogenetische Untersuchungen an knotigen Veränderungen des menschlichen Schilddrüse. *Schweiz Med Wschr, 100*:633, 1970.

Hsu, T. C.: Mammalian chromosomes *in vitro*. IV. Some human neoplasms. *J Nat Cancer Inst, 14*:905, 1954.

Huang, C. S.; Imamura, T.; Moore, G. E.: Chromosomes and cloning efficiencies of hematopoietic cell lines derived from patients with leukemia, melanoma, myeloma and Burkitt lymphoma. *J Nat Cancer Inst, 43*:1129, 1969.

Ishihara, T.; Moore, G. E.; Sandberg, A. A.: Chromosome constitution of cells in effusions of cancer patients. *J Nat Cancer Inst, 27*:893, 1961.

Ishihara, T.; Moore, G.; Sandberg, A. A.: The *in vitro* chromosome constitution of cells from human tumors. *Cancer Res, 22*:375, 1962.

Ishihara, T.; Kikuchi, Y.; Sandberg, A. A.: Chromosomes of twenty cancer effusions: Correlation of karyotypic, clinical, and pathologic aspects. *J Nat Cancer Inst, 30*:1303, 1963.

Ishihara, T.; Makino, S.; Tonomura, A.: Cytological studies of tumors, XXVII. The chromosomes of thirty human tumors. *Z Krebsforsch, 63*:184, 1959.

Ishii, S.: Chromosome studies of human bone tumors *in vivo* and *in vitro*. *Gann, 56*:251, 1965.

Laurent, M.; Rousseau, M. F.; Nezelof, C.: Etude caryotypique d'un tératome sacro-coccygien. *Ann Anat Path, 13*:413, 1968.

Kong-oo Goh: Large abnormal acrocentric chromosome associated with human malignancies. Possible mechanism of establishing clone of cells. *Arch Int Med, 122*:241, 1968.

Miles, C. P.: Chromosomal alterations in cancer. *Med Clin North Amer, 50*:875, 1966.

Miles, C. P.: Chromosome analysis of solid tumors. I. Twenty-eight nonepithelial tumors. *Cancer, 20*:1253, 1967.

Miles, C. P.: Chromosome analysis of solid tumors. II. Twenty-six epithelial tumors. *Cancer, 20*:1274, 1967.

Miles, C. P., and Gallagher, R. E.: Chromosomes of a metastatic human cancer. *Lancet, 2*:1145, 1961.

Sandberg, A. A., and Yamada, K.: Chromosomes and causation of human cancer and leukemia. I. Karyotype diversity in a single cancer. *Cancer, 19*:1869, 1966.

Sandberg, A. A.; Yamada, K.; Kikuchi, Y.; Takagi, N.: Chromosomes and causation of human cancer and leukemia. III. Karyotypes of cancerous effusions. *Cancer, 20*:1099, 1967.

Socolow, E.; Engel, E.; Mantooth, L.; Stanbury, J.: Chromosomes of human thyroid tumors. *Cytogenetics, 3*:394, 1964.

Spiers, A. S. D., and Baikie, A. G.: Reticulum cell sarcoma: demonstration of chromosomal changes analogous to those in SV 40 transformed cells. *Br J Cancer, 21*:679, 1967.

Steele, H. D.; Monocha, S. L.; Stich, H. F.: Desoxyribonucleic acid content of epidermal in-situ carcinomas. *Br Med J, 2*:1314, 1963.

Stich, H. F., and Steele, H. D.: DNA content of tumor cells. III. Mosaic composition of sarcomas and carcinomas in man. *J Nat Cancer Inst, 28*:1207, 1962.

Vincent, P. C.; Vandenburg, R. A.; Neate, R.; Nicholls, A.: Chromosome analysis in the diagnosis of malignant effusions: report of a case. *Med J Aust, 1*:155, 1964.

White, L., and Cox, D.: Chromosome changes in a rhabdomyosarcoma during recurrence and in cell culture. *Br J Cancer, 21*:684, 1967.

INDEX

A

Acute aleukemic leukemia, 86, 87
 benzene leukemia, 84-86
 eosinophilic leukemia, 86, 87
 erythroleukemia, 62, 96, 97
 granulocytic leukemia (see acute myeloblastic leukemia)
 lymphoblastic leukemia, 20, 24, 26, 42, 82
 monoblastic leukemia, 86
 myeloblastic leukemia, 21, 24, 80-82- 107
 myelomonocytic leukemia, 82, 183
 myeloid leukemia (see acute myeloblastic leukemia)
 promyelocytic leukemia, 86
 smoldering leukemia, 107
Adenoma, colonic, 154-156
Agammaglobulinemia, Bruton's type, 26
Aleukemic leukemia, 86, 87
Alkaline phosphatase (see leukocyte alkaline phosphatase)
Amphetamines, 34
Anemia, aplastic, 11, 106-109
 idiopathic, 107, 108
 megaloblastic, 77
 pernicious, 107, 108
Anemia, pernicious, 107, 108
 refractory, 107
 sideroachrestic, 11, 107, 108, 124
 sideroblastic, 95, 107
Anus carcinoma, 161
Arrhenoblastoma, 149-152
Astrocytoma, 177, 178
Ataxia-telangiectasia, 26
Atypia, cervix uteri, 133, 134, 137
Au^{198}, 160

B

Banding pattern of chromosomes, 46-49

Barr-bodies (see sex chromatin)
Benzene (leukemia), 34, 84, 86
Bladder (urinary) carcinoma, 186, 187, 188, 189
 (urinary) papilloma, 186, 189
Blastic crisis of chronic myeloid leukemia, 63-71
Bloom's syndrome, 26, 109
Bovine, leukemia, 41
Bowen's disease, 10
Break, chromosome, 7, 26, 32, 34, 171, 183
Breast carcinoma, 23, 131-133, 140, 141
Bronchial adenoma, 171
 carcinoma, 171-174
Buccal mucosa cancer, 157
Burkitt's lymphoma, 33, 34, 47, 119, 122

C

Canine veneral sarcoma, 41
Carzinophilin, 159, 160
Cerebellar sarcoma, 182
Cervix uteri carcinoma, 15, 131, 133-137, 174
 uteri dysplasia (see dysplasia)
Chronic lymphoid leukemia, 15, 16, 24, 27, 55-77, 124, 125
 myeloid leukemia, 11, 14, 16, 20, 24, 25, 27, 55-77
Chédiak-Higashi syndrome, 26, 119
Chicken pox virus, 32
Childhood myeloid leukemia, 15
Chondrosarcoma, 26, 195
Chordoma, 176
Choriocarcinoma testis,, 149-152
Chorionepithelioma, 139
Chorionic villi, 139
Christchurch chromosome (Ch$_1$), 15, 124
Clastogen, 34

Clonal evolution, 9, 20, 26, 37
Colonic carcinoma, 161-165
Common pathway of chromosomal evolution, 16, 165
Congenital leukemia, 20
Constitutive heterochromatin (see heterochromatin)
Corpus uteri, carcinoma, 137, 138
Cri-du-chat syndrome, 21
Cystadenocarcinoma, 131
Cystadenoma, ovarian, 131, 132
Cystic hyperplasia, 133
   mammary disease, 133

D

Deletion, 8, 14, 15
Dermoid cyst, 131
DiGuglielmo syndrome, 96
DNA, main band, 46
   satellite repetitive, 17, 46, 50, 195
Double minute chromosomes (see minute chromosomes)
Down's syndrome, trisomy 21, 11, 19-25, 58, 59, 179, 181
Dq-syndrome, 179, 181, 182
Duplication of Philadelphia chromosome, 63-65
Dysgerminoma, 149-152
Dysplasia, cervix uteri, 133, 134, 135, 137

E

E17qi chromosome, 68
E18p- chromosome, 119, 122
Enchondromatosis, 26
Endometrial carcinoma, 138
   hyperplasia, atypical, 138
Endometrium, polyploidy in, 7
Endoreduplication, 7, 8, 159, 163
Eosinophilic leukemia, 86, 87
Ependymoma, 178
Epiglottis neoplasm, 171
Epstein-Barr virus, 33
Erythremic myelosis, 96
Erythroleukemia, 62, 96-98
Esophagus carcinoma, 157
Established cell lines, 14
Evolution, clonal (see clonal evolution)

F

Facultative heterochromatin (see heterochromatin and sex chromatin)
Familial cancer, 22-25, 26, 27, 122, 182
   lymphoma, 122
Fanconi's anemia, 25, 26, 106-109
Fibromyoma uteri, 138
Fibrosarcoma, 195
Fission, centromere, 39
Fluorescent staining, 46, 47, 58, 59
Folate deficiency anemia, 107, 108
Follicular lymphoma, 119
Foot-and-mouth disease, 32
Forbidden combination, 10
Fp- marker, 95, 107

G

Gammopathies, 106
Gap, chromosome, 7, 26, 34
Gardner's syndrome (see polyposis of the colon, familial)
Gastric cancer, 158
Glioma, 26, 176-177, 178
Glucose-6-phosphate dehydrogenase,
Goiter, 194
Gonadal dysgenesis, 22
Gonadoblastoma, 149-152
Granulocytic hyperplasia, 101

H

Haematological non-leukemic disorders, 106-118
Hepatic carcinoma, 165-166
Hepatitis, infectious, 32
Herpes simplex virus, 32
   zoster virus, 32
Heterochromatin, constitutive, 46, 195
   facultative, 46, 140, 141, 150, 171, 174
Hodgkin's disease, 119-122
Hydatidiform mole, 139, 140
Hypophyseal tumors, 176

I

Idiopathic anemia, 107, 108

Immunologic deficiencies, 119
  rejection, 10
In situ, carcinoma of cervix, 133, 134, 135, 137
Insertion, 8
Inversion, 8
Isochromosome, 7, 8, 68

K

Klinefelter's syndrome, 22, 23, 24

L

Lactational hyperplasia, 133
Lagging, chromosome, 8
Larynx carcinoma, 171
Leiomyosarcoma, colonic, 160
  retroperitoneal, 166, 195
Leukemia *(see* acute leukemia, chronic leukemia)
Leukemia, congenital, 20
Leukemoid reactions, 93
Leukocyte alkaline phosphatase, 16, 27, 59
Liposarcoma, 195
Liver, polyploidy in, 7
  cancer *(see* hepatic carcinoma)
LSD and Ph_1 chromosome, 34
Lung carcinoma, 23, 170-174
Lymphoma, malignant, 15, 26, 119, 122
Lymphosarcoma, 23, 26, 119, 195

M

Macroglobulinemia Waldenström *(see* Waldenström's m.)
Madison chromosome,
Marihuana, 34
Marker chromosome, 6, 7, 8, 9, 11, 15, 16, 23, 27, 68, 95, 106, 107, 110-113, 119-122, 124, 150-152
Martineau marker 150-152
Maxilla cancer, 57
Measles virus, 32
Medulloblastoma, 177-178,184
Megakaryocytes, 7, 57, 58, 108
Megaloblastic anemia, 77, 97
Melanoma, 192, 193, 194
Melbourne chromosome, 15, 122

Meningeal sarcoma, 182
Meningioma, 176, 182
Meningitis, aseptic, 32
Mescaline, 34
MG chromosome, 106
Microchromosomes *(see* minute chromosomes)
Microspectrophotometry, 154, 157, 158, 159, 187, 194
Minute chromatin bodies *(see* minute chromosomes)
  chromosomes, 82, 171, 177, 178, 179, 180, 181, 182, 183, 184
Monosomy G, 82
Multiple myeloma, 106, 113-115
Multipolar mitosis, 6
Mumps virus, 32
Myelodysplasia, 106
Myelofibrosis, 99, 100
Myeloid metaplasia, 99, 100
Myeloproliferative disorders, 92-105

N

Nephroblastoma, 189
Neuroblastoma, 179, 180, 181, 183, 184
Neurofibromatosis, 26, 182
Nondisjunction, 7
Normal karyotype (in cancer), 14, 33, 40, 62, 80, 81, 87, 133, 134, 135, 138, 139, 151, 155, 165, 166, 180, 189, 195
Nucleolus, 8

O

Oligodendroglioma, 177
Oncocytoma of thyroid *(see* thyroid)
Oral carcinoma, 23, 156-158
Osteosarcoma, 195
Ovary carcinoma, 15, 23, 131, 132

P

Paget's disease, 10
Pancreas carcinoma, 165-166
Pancytopenia, 106, 108
Penis carcinoma, 152
Pernicious anemia, 107, 108
Peritoneum carcinoma, 165-166, 195

Pharynx neoplasm, 171
Philadelphia (Ph$_1$) chromosome, 11, 15, 16, 29, 25, 27, 34, 56-62, 81, 82
Phytohaemagglutinin, 58, 121, 124
Point mutation, 14, 16, 87, 88
Polycythemia vera, 11, 27, 62, 93-96
Polymorphonuclear leukocytes, 59
Polyoma virus, 32, 51
Polyposis of the colon, familial, 10, 26, 154, 155, 163
Preleukemias, 106
Premalignancy, 10, 106, 133, 154-156
Promyelocytic leukemia, 86
Prostate carcinoma, 189
Pseudohermaphroditism, 149
Pseudostemline concept, 43
Pulverization of chromosomes, 32

R

Reassociating DNA, 46
Rectum cancer (see anus carcinoma)
Reed-Sternberg cells, 119-121
Refractory anemia, 107
Regulator genes, 17, 20
Repetitive DNA (see DNA)
Repressing genes, 17, 20
Reticulosarcoma, 124, 195
Retinoblastoma, 179, 181, 182
Rhabdomyosarcoma, 183, 188, 195
Ring chromosome, 7, 8, 27, 98, 133, 163, 166, 170, 183, 187
Robertsonian fusion and fission, 38-41
Rous sarcoma virus, 32

S

Satellite DNA (see DNA)
Seminoma, 149-152
Sendai virus, 32
Sex chromatin, 140, 141, 150, 171, 174
Sideroachrestic anemia (see anemia)
Sideroblastic anemia (see anemia)
Simian virus 40 (SV40), 50
Speciation, chromosome evolution, 37-39, 43, 44
Specific common pathway (see common pathway of chromosomal evolution)
Spherocytosis, hereditary, 34
Spleen cancer, 195
Stemline, 6, 16, 26, 78, 79
Stimulator genes, 17
Stomach cancer (see gastric cancer)
Structural genes, 20
Submandibular gland carcinoma, 157

T

Teratoblastoma of testis, 149-152
Teratoma, ovarian, 132
Teratoma, sacrococcygeal, 195
Testicular tumors, 10, 15, 149
Therapeutics, 70, 71
Thrombocythemia, 11, 62, 93, 98
Thrombocytosis, 98
Thyroid adenoma, 194
  carcinoma, 192, 194
  fibrosis, 194
  oncocytoma, 194
Tonsils neoplasm, 171
Tongue carcinoma, 157
Toyomycin, 160
Translocations, 8, 14, 22-25
Translocation D/D, 22, 23, 25
  D/G, 24
Triple X syndrome (see
Trisomy 21, (see Down's syndrome)
  D, 21
  E, 21, 22
  F, 22
Trophoblastic tumors, 139
Tuberous sclerosis, 26
Turner's syndrome (XO), 22, 149
Twins, 25, 26, 59, 79, 83

U

Urethra carcinoma, 189, 190
Ureter carcinoma, 189, 190
Urinary bladder, polyploidy, 7

V

Veneral canine sarcoma (see canine veneral sarcoma)
Vincristine, 170

W

Waldenström's macroglobulinemia, 15, 25, 27, 106, 110-113
W-chromosome, 106, 110-113
Wiskott-Aldrich syndrome, 119

Y

Y-chromosome, loss of, 70, 109
Yellow fever vaccine, 32

X

XO syndrome (Turner), 22, 149
XXX syndrome, 23, 25
XXY syndrome (Klienefelter), 22, 23, 24